Software-Defined Networking

To my parents, my sisters and my wonderful readers
Fetia Bannour

To my beloved family, my wife and my dear children
Sami Souihi

To my oldest, beloved and magnificent daughter Ikram
on her birthday this year.
Abdelhamid Mellouk

New Generation Networks Set

coordinated by
Abdelhamid Mellouk

Volume 2

Software-Defined Networking

*Extending SDN Control
to Large-Scale Networks*

Fetia Bannour
Sami Souihi
Abdelhamid Mellouk

WILEY

First published 2022 in Great Britain and the United States by ISTE Ltd and John Wiley & Sons, Inc.

ISTE Ltd
27-37 St George's Road
London SW19 4EU
UK

www.iste.co.uk

John Wiley & Sons, Inc.
111 River Street
Hoboken, NJ 07030
USA

www.wiley.com

Library of Congress Control Number: 2022941880

British Library Cataloguing-in-Publication Data
A CIP record for this book is available from the British Library
ISBN 978-1-78630-849-8

Contents

Acronyms

API	Application Programming Interface
AS	Autonomous System
CAP	Consistency Availability Performance
CDN	Content Delivery Network
CLARA	Clustering LARge Applications
CPP	Controller Placement Problem
DDBS	Distributed DataBase System
DHT	Distributed Hash Table
DoS	Denial-of-Service
ForCES	Forwarding and Control Element Separation
FSM	Finite-State Machine
IETF	Internet Engineering Task Force

IoT	Internet of Things
IXP	Internet eXchange Point
MD-SAL	Model-Driven Service Abstraction Layer
ML	Machine Learning
MOCO	Multi-Objective Combinatorial Optimization
NIB	Network Information Base
NSGA-II	Non-dominated Sorting Genetic Algorithm II
ODL	OpenDayLight
OF	OpenFlow
ONF	Open Networking Foundation
ONOS	Open Network Operating System
PACELC	Partition, tradeoff Availability and Consistency, Else, tradeoff Latency and Consistency
PAM	Partitioning Around Medoids
POCO	Pareto-Optimal COntroller
PSA	Pareto Simulated Annealing
QL	Q-Learning
QoE	Quality of Experience

QoS   Quality of Service

RL   Reinforcement Learning

RSM   Replicated State Machine

SDN   Software-Defined Networking

SDX   Software-Defined eXchange

SLA   Service-Level Agreement

SPOF   Single Point of Failure

TE   Traffic Engineering

UDP   User Datagram Protocol

WAN   Wide Area Network

XFSM   eXtended Finite-State Machine

Preface

Due to the emergence of new kinds of communication and networking technologies (e.g. the Internet of Things (IoT), mobile trends, network virtualization) and the rise of many advanced services (e.g. real-time services, e-health, multimedia, smart cities, gaming) supported by these technologies, today's networks – considered relatively static, "ossified" and "challenging to manage"– are no longer suitable to handle the complexity and diversity of network information being disseminated in today's modern and dynamic networking environments.

There is a strong need to shift the current network architecture to a new model that adapts to such changes and leverages new control strategies to ease network management and automation, leading to better network performance and lower operating costs. In this context, software-defined networking (SDN) has emerged as a new networking paradigm that decouples network control and forwarding functions, enabling the network control to become directly programmable and the underlying infrastructure to be abstracted for applications and services.

SDN attempts to centralize the network control, thus offering improved visibility and flexibility to manage the network, optimize its performance and reduce its operating costs. However, centralized SDN designs, in which the control plane logic is physically centralized in a single software component called the SDN controller, present numerous challenges including the issues of control plane reliability, scalability and performance. To meet these challenges, it is necessary for the SDN control architecture to evolve toward a physically decentralized system. However, such physically distributed but logically centralized SDN platforms bring additional challenges.

In this book, we aim to provide a thorough exploration of the SDN technology and, more importantly, we deal with the SDN decentralization problem in the context of large-scale networks. We propose novel approaches to decentralize the SDN control plane without forgoing the centralization benefits of SDN. Part of this book was initially based on the work conducted within the framework of Fetia Bannour's PhD thesis. This work was subsequently developed into a book to facilitate understanding of the decentralized SDN control plane. The latter may indeed be implemented using the existing distributed SDN controllers. However, their significant number, along with their particular pros and cons, made the choice extremely difficult for those who attempted to adopt a distributed SDN architecture in large-scale deployments.

To provide useful guidelines for such SDN research and deployment initiatives, this book reviews the SDN concept by studying the SDN architecture compared to the traditional one and provides a detailed analysis of state-of-the-art distributed SDN controller platforms by assessing their advantages and drawbacks, classifying them in novel ways (physical and logical classifications) and comparing them with respect to various criteria. Additionally, a thorough discussion on the major challenges of existing distributed SDN controller platforms is provided along with insights into emerging and future trends in that area. Furthermore, to tackle some of the most prominent challenges related to the decentralization of the SDN control plane in large-scale networks, three novel approaches are proposed in this book.

The first approach addresses the SDN controller placement problem by proposing scalability and reliability aware strategies for the placement of the SDN controllers at scale, with respect to multiple reliability and performance criteria and according to different uses and contexts. These strategies use different types of multi-criteria optimization algorithms, which are compared in terms of computation time, and the quality of final controller placement configurations.

The second and third approaches investigate the knowledge sharing problem in a distributed SDN cluster by proposing adaptive and continuous consistency models. The main aim of these two approaches is to achieve a consistency adaptation strategy that provides balanced trade-offs at runtime between the application's continuous performance and consistency requirements. These real-time trade-offs should provide minimal application inter-controller overhead while satisfying the application-defined thresholds specified in the application's service-level agreements (SLAs). These models primarily focus on the anti-entropy reconciliation mechanisms. Then, they address the replication mechanisms by proposing an intelligent Quorum replication strategy. These approaches were validated using two SDN applications with eventual consistency needs that are developed on top of the open-source Open Network Operating System (ONOS) controllers: a source routing application and a CDN-like application.

When writing this book, we were mainly driven by our belief that SDN, together with network virtualization, will play a significant role in enabling a "full network softwarization" and reshaping the next generation of computer networks. We were also motivated by the lack of research work on decentralized SDN, which extends SDN control to large-scale networks. Our purpose is to provide useful guidelines and lessons learned for dealing with the decentralization problem in SDN for academic and industrial research purposes.

This book is a start but also leaves many questions unanswered. We hope that it will inspire the new generation of researchers. It would not have been possible without the valuable support of our students and colleagues, whom the authors would like to thank warmly.

Finally, the authors hope the readers will enjoy reading this book and learn many useful ideas and overviews for their own work and studies.

August 2022

Fetia BANNOUR
École Nationale Supérieure d'Informatique
pour l'Industrie et l'Entreprise (ENSIIE)

Sami SOUIHI
Université Paris-Est Créteil (UPEC)

Abdelhamid MELLOUK
Université Paris-Est Créteil (UPEC)

Introduction

I.1. General context

The unprecedented growth in demands and data traffic and the emergence of network virtualization, along with the ever-expanding use of mobile equipment in the modern network environment, have highlighted major problems that are essentially inherent to the Internet's conventional architecture. This made the task of managing and controlling the information coming from a growing number of connected devices increasingly complex and specialized.

Indeed, the traditional networking infrastructure is considered highly rigid and static as it was initially conceived for a particular type of traffic, namely monotonous text-based contents, which makes it poorly suited to today's interactive and dynamic multimedia streams, generated by increasingly demanding users. Along with multimedia trends, the recent emergence of the Internet of Things (IoT) has allowed for the creation of new advanced services with more stringent communication requirements in order to support its innovative use cases. In particular, e-health is a typical IoT use case where the healthcare services delivered to remote patients (e.g. diagnosis, surgery, medical records) are highly intolerant of delay, quality and privacy. Such sensitive data and life-critical traffic are barely supported by traditional networks.

Furthermore, in the traditional architecture where the control logic is purely distributed and localized, solving a specific networking problem or adjusting a particular network policy requires acting separately on the affected devices and manually changing their configuration. In this context, the current growth in devices and data has exacerbated scalability concerns by making such human interventions and network operations harder and more error prone.

Overall, it has become particularly challenging for today's networks to deliver the required level of quality of service (QoS), let alone the quality of experience (QoE) that introduces additional user-centric requirements. To be more specific, relying solely on the traditional QoS, based on technical performance parameters (e.g. bandwidth and latency), turns out to be insufficient for today's advanced and expanding networks. Additionally, meeting this growing number of performance metrics is a complex optimization task, which can be treated as an NP-complete problem. Alternatively, network operators are increasingly realizing that the end-user's overall experience and subjective perception of the delivered services are as important as QoS-based mechanisms. As a result, current trends in network management are heading toward this new concept, commonly referred to as the QoE, to represent the overall quality of a network service from an end-user perspective.

That said, this huge gap between, on the one hand, the advances achieved in both computer and software technologies and, on the other hand, the traditional non-evolving and *hard to manage* (Kreutz et al. 2015; Bannour et al. 2018b) underlying network infrastructure supporting these changes has stressed the need for an automated networking platform (Samaan and Karmouch 2009) that facilitates network operations and matches today's network requirements such as the IoT needs (Ren et al. 2019; Montaño et al. 2021). In this context, several research strategies have been proposed to integrate automatic and adaptive approaches into the current infrastructure for the purpose of meeting the challenges of scalability, reliability and availability for real-time traffic, and therefore guaranteeing the user's QoE.

While radical alternatives argue that a brand new network architecture should be built from scratch by breaking with the conventional network architecture and bringing fundamental changes to keep up with current and future requirements, other realistic alternatives are favored for introducing slight changes tailored to specific needs and for making a gradual network architecture transition, without causing costly disruptions to existing network operations.

In particular, the early overlay network alternative introduces an application layer overlay on top of the conventional routing substrate to facilitate the implementation of new network control approaches. However, the obvious disadvantage of overlay networks is that they depend on several aspects (e.g. selected overlay nodes) to achieve the required performance. Moreover, such networks can be criticized for compounding the complexity of existing networks due to the additional virtual layers.

On the other hand, the recent software-defined networking (SDN) paradigm (Feamster et al. 2014) offers the potential to program the network and thus facilitates the introduction of automatic and adaptive control approaches by separating hardware (data plane) and software (control plane), enabling their independent evolution. SDN aims for the centralization of the network control, offering improved visibility and better flexibility to manage the network and optimize its performance. When compared to the overlay network alternative, SDN has the ability to control the entire network, not just a selected set of nodes, and use a public network for transporting data. Moreover, SDN spares network operators the tedious task of temporarily creating the appropriate overlay network for a specific use case. Instead, it provides an inherent programmatic framework for hosting control and security applications that are developed in a centralized way while taking into consideration the IoT requirements (Li et al. 2016; Ren et al. 2019; Montaño et al. 2021) to guarantee the user's QoE.

I.2. Problem statement and motivations

Despite the considerable interest in SDN, its deployment in the industrial context is still in the relatively early stages. Indeed, there may be a long road ahead before technology matures and standardization efforts pay off so that the full potential of SDN can be achieved.

In fact, along with the hype and excitement, there have been several concerns and questions regarding the widespread adoption of SDN networks. For instance, research studies on the feasibility of the SDN deployment have revealed that the physical centralization of the control plane in a single programmable software component, called the controller, is constrained by several limitations such as the issues of scalability, availability and reliability. Gradually, it became unavoidable to think about the control plane as a distributed system (Canini et al. 2014; Sarmiento et al. 2021), in which several SDN controllers are in charge of handling the whole network while maintaining a logically centralized network view.

In this respect, networking communities argued about the best way to implement distributed SDN architectures while taking into account the new challenges brought by such distributed systems. Consequently, several SDN solutions have been explored and many SDN projects have emerged. Each proposed SDN controller platform adopted a specific architectural design approach based on various factors such as the aspects of interest, the performance goals, the deployed SDN use case and also the trade-offs involved in the presence of multiple conflicting and competing challenges.

Here, we underline the importance of conducting a thorough analysis of the proposed SDN solutions to envisage the potential trends that may drive future research in the area. In particular, we place a special focus on distributed SDN control designs with the aim of solving some of the major challenges encountered in the decentralization of the SDN control planes in the context of large-scale deployments.

The main motivations for this work are as follows:

– ensuring a thorough understanding of existing state-of-the-art distributed SDN controller platforms, and developing a critical awareness of the ongoing and future key research and operational challenges facing the design and deployment of such platforms;

– proposing novel approaches for decentralizing the SDN control plane in large-scale networks. Such a decentralized SDN control plane should be efficient (i.e. scalable, high-performance and robust) as it should meet the SDN controller application requirements (e.g. scalability, reliability and consistency);

– paving the way for the emergence of a new common standard for the distributed SDN control plane. This standard should also ensure the inter-controller communication between different vendor-specific controller technologies (i.e. the interoperability challenge).

I.3. Main contributions

In this section, we outline the main contributions of this work. More specifically, we propose novel approaches for decentralizing the software-defined networking (SDN) control plane in large-scale networks, while tackling some of the major associated challenges:

1) *scalability* and *reliability* aware strategies for the *placement of distributed SDN controllers* at scale using different types of multi-criteria optimization algorithms (see Chapter 3);

2) an adaptive and continuous *consistency* model for the distributed SDN controllers: a novel anti-entropy reconciliation mechanism for applications (with eventual consistency needs) on top of the ONOS controllers (see Chapter 4);

3) an adaptive and continuous *consistency* model for the distributed SDN controllers: a novel Quorum-based replication strategy for applications (with eventual consistency needs) on top of the ONOS controllers (see Chapter 5).

Additionally, given the lack of available literature on the subject of decentralized SDN control and given its relevance today, our work also provides the following:

– a survey on distributed control in SDN: an overview and taxonomy of current SDN controller platforms (i.e. a physical classification and a logical classification) (see Chapter 1);

– a thorough analysis of the challenges encountered by the discussed state-of-the-art distributed SDN controller platforms, and the different approaches adopted for solving these challenges (see Chapter 2).

I.4. Structure of the book

Chapter 1 presents a survey on SDN with a special focus on distributed SDN control solutions. In addition to explaining the fundamental elements of the SDN architecture, this chapter proposes a taxonomy of the most prominent state-of-the-art SDN controller platforms by classifying them in two different ways: a physical classification and a logical classification.

Chapter 2 provides a thorough analysis of the major open challenges faced by the existing distributed SDN controller platforms discussed in the previous chapter. These challenges include the issues of scalability, reliability, consistency and interoperability of the SDN control plane. Furthermore, this chapter explores potential approaches to tackle these challenges for an optimal SDN deployment and provides some useful insights into the emerging and future trends in the design of efficient distributed SDN control planes.

Chapter 3 addresses the distributed SDN control problem by tackling the SDN controller placement problem in large-scale IoT-like networks. It puts forward novel scalability and reliability aware controller placement strategies that deal with several aspects of the controller placement optimization problem with respect to multiple reliability and performance criteria and according to different uses and contexts. These strategies use two different types of heuristic-based algorithms: a clustering algorithm based on PAM and a modified genetic algorithm called NSGA-II. These multi-criteria algorithms are compared in terms of computation time, as well as the quality of final controller placement configurations.

Chapter 4 focuses on the distributed SDN control problem by tackling the knowledge sharing problem between the distributed SDN controllers. It proposes an adaptive multi-level consistency model following the concept of continuous consistency for the distributed SDN controllers. This approach is implemented for a source routing application on top of the open-source ONOS controllers. It involves turning ONOS's eventual consistency model into an adaptive consistency model using the *anti-entropy reconciliation period* as a control knob for an adaptive fine-grained tuning of consistency levels. Our proposed consistency strategy is aimed at ensuring the application's continuous consistency requirements (i.e. numerical error bounds), as specified in the given application service-level agreement (SLA). Its purpose is also to minimize the anti-entropy reconciliation overhead as compared to ONOS's static consistency scheme at scale.

Chapter 5 further addresses the knowledge sharing problem in the distributed SDN control by proposing an adaptive and continuous consistency model for the distributed ONOS controllers. The approach is implemented for a CDN-like application on top of ONOS. It consists of changing ONOS's eventual consistency model to an adaptive consistency model by turning ONOS's optimistic replication technique into a more scalable replication strategy following Quorum-replicated consistency. The main focus is placed on improving ONOS's replication mechanism: it uses the *read and write Quorum parameters* as adjustable control knobs for a fine-grained consistency tuning rather than relying on anti-entropy reconciliation mechanisms (Chapter 4). The main objective is to find optimal partial Quorum configurations at runtime that achieve, under changing network and workload conditions, balanced trade-offs between the application's continuous performance (latency) and consistency (staleness) requirements. These real-time trade-offs should provide minimal application inter-controller overhead while satisfying the application-defined thresholds specified in the given application SLA.

Conclusions and Perspectives is the final chapter of the book and gives an insight into our ongoing and future work, and perspectives in the area of distributed SDN control.

Toward a Decentralized SDN Control Architecture: Overview and Taxonomy

1.1. Introduction

In contrast to the decentralized control logic that underpins the construction of the Internet as a complex bundle of box-centric protocols and vertically integrated solutions, the software-defined networking (SDN) paradigm advocates the separation of the control logic from hardware and its centralization in software-based controllers. These key tenets offer new opportunities to introduce innovative applications and incorporate automatic and adaptive control aspects, thereby easing network management and guaranteeing the user's QoE.

However, despite the interest surrounding SDN, adoption raises many challenges, including the scalability and reliability issues of centralized designs that can be addressed with the physical decentralization of the control plane. However, such physically distributed but logically centralized systems bring an additional set of challenges.

This chapter presents a survey on SDN with a special focus on distributed SDN control. In section 1.2, we start by describing the promises and solutions offered by SDN compared to conventional networking. We also elaborate on the fundamental elements of the SDN architecture.

Then, we expand our knowledge of the different approaches to SDN by exploring the wide variety of existing SDN controller platforms. In doing so, we intend to place a special emphasis on distributed SDN solutions and classify them in two different ways. In section 1.3, we propose a physical classification of state-of-the-art SDN control plane architectures into centralized and distributed (flat or hierarchical) in order to highlight the SDN performance, scalability and reliability challenges. In section 1.4, we put forward a logical classification of distributed SDN control plane architectures, logically centralized and logically distributed, while tackling the associated state consistency and knowledge dissemination issues.

1.2. Software-defined networking: a centralized control architecture

1.2.1. *Conventional networking and the SDN paradigm*

Over the past few years, the need for a new approach to networking has been expressed to overcome the many issues associated with current networks. In particular, the main vision of the SDN approach is to simplify networking operations, optimize network management and introduce innovation and flexibility compared to legacy networking architectures.

In this context, and in line with the vision of Kim and Feamster (2013), four key reasons for the problems encountered in the management of existing networks can be identified:

1) *Complex and low-level network configuration*: network configuration is a complex distributed task in which each device is typically configured in a low-level vendor-specific manner. Additionally, the rapid growth of the network, together with the changing networking conditions, have resulted in network operators constantly performing manual changes to network configurations, thereby compounding the complexity of the configuration process and introducing additional configuration errors.

2) *Dynamic network state*: networks are growing dramatically in size, complexity and consequently in dynamicity. Furthermore, with the rise of mobile computing trends as well as the advent of network virtualization (Bari et al. 2013; Alam et al. 2020) and cloud computing (Zhang et al. 2010; Sharkh et al. 2013; Shamshirband et al. 2020), the networking environment becomes even more dynamic as hosts are continually moving, arriving and departing due to the flexibility offered by VM migration, thus making traffic patterns and network conditions change in a more rapid and significant way.

3) *Exposed complexity*: in today's large-scale networks, network management tasks are challenged by the significant complexity presented by distributed low-level network configuration interfaces. This complexity is mainly generated by the tight coupling between the management, control and data planes, where many control and management features are implemented in hardware.

4) *Heterogeneous network devices*: current networks consist of a large number of heterogeneous network devices including routers, switches and a wide variety of specialized middle-boxes. Each of these appliances has its own proprietary configuration tools and operates according to specific protocols, therefore increasing both complexity and inefficiency in network management.

As a result, network management is becoming more difficult and challenging given that the static and inflexible architecture of legacy networks is ill-suited to cope with today's increasingly dynamic networking trends, and to meet the QoE requirements of modern users. This has fueled the need for the enforcement of complex, high-level policies to adapt to current networking environments, and for the automation of network operations to reduce the tedious workload of low-level device configuration tasks.

In this sense, and to deliver the goals of easing network management in real networks, operators considered running dynamic scripts as a way to automate network configuration settings before realizing the limitations of such approaches, which led to misconfiguration issues. It is worth noting, however, that recent approaches to scripting configurations and network automation are becoming relevant (e.g. Ansible).

The SDN initiative led by the Open Networking Foundation (ONF), on the other hand, proposes a new open architecture to address current networking challenges with the potential to facilitate the automation of network configurations and, better yet, fully program the network. Unlike the conventional distributed network architecture (Figure 1.1(a)) in which network devices are closed and vertically integrated, bundling software with hardware, the SDN architecture (Figure 1.1(b)) raises the level of abstraction by separating the network data and control planes. In this way, network devices become simple forwarding switches; all the control logic is centralized in software controllers, providing a flexible programming framework for the development of specialized applications and deployment of new services.

Such aspects of SDN are believed to simplify and improve network management by offering the possibility to innovate, customize behaviors and control the network according to high-level policies expressed as centralized programs. This therefore bypasses the complexity of low-level network details and overcomes the fundamental architectural problems raised in points 1) and 3). Added to these features is the ability of SDN to easily cope with the heterogeneity of the underlying infrastructure (outlined in point 4)) thanks to the SDN southbound interface abstraction.

More detailed information on the SDN-based architecture, which is split vertically into three layers (see Figure 1.2), is provided in the next section.

1.2.2. *The SDN architecture*

The SDN-based architecture is split vertically into three layers (see Figure 1.2). Detailed information about the SDN architecture is provided in the sections that follow.

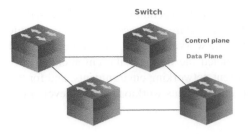

a) Traditional architecture

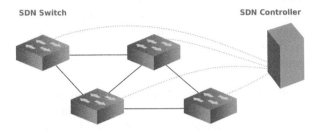

b) SDN architecture

Figure 1.1. *Conventional networking versus software-defined networking. For a color version of this figure, see www.iste.co.uk/bannour/software.zip*

1.2.2.1. *SDN data plane*

The data plane, also known as the forwarding plane, consists of a distributed set of forwarding network elements (mainly *switches*) in charge of forwarding packets. In the context of SDN, the control-to-data plane separation feature requires the data plane to be remotely accessible for software-based control via an open vendor-agnostic southbound interface.

Both OpenFlow (McKeown et al. 2008; Costa et al. 2021) and ForCES (Forwarding and Control Element Separation) (Doria et al. 2010; Anerousis et al. 2021) are well-known candidate protocols for the southbound interface. They both follow the basic principle of splitting the control plane and the forwarding plane in network elements, and they both standardize the communication between the two planes. However, these solutions are different in many aspects, especially in terms of network architecture design.

Standardized by IETF, ForCES (Doria et al. 2010; Anerousis et al. 2021) introduced separation between the control plane and the forwarding plane. In doing so, ForCES defines two logic entities that are logically kept in the same physical device: the control element (CE) and the forwarding element (FE). However, despite being a mature standard solution, the ForCES alternative did not gain widespread adoption by major router vendors.

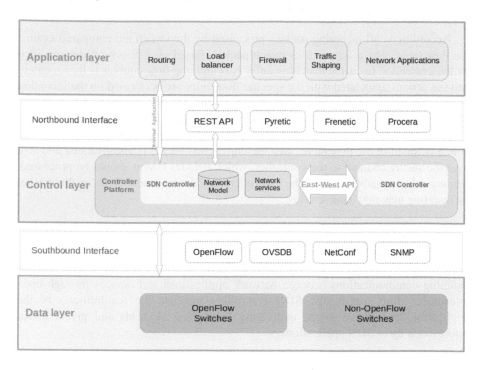

Figure 1.2. *A three-layer distributed SDN architecture. For a color version of this figure, see www.iste.co.uk/bannour/software.zip*

On the other hand, OpenFlow (McKeown et al. 2008; Costa et al. 2021) received major attention in both the research community and industry. Standardized by the ONF, it is considered the first widely accepted communication protocol for the SDN southbound interface.

OpenFlow enables the control plane to specify, in a centralized way, the desired forwarding behavior of the data plane. Such traffic forwarding decisions reflect the specified network control policies and are translated by *controllers* into actual packet forwarding rules, populated in the flow tables of OpenFlow *switches*.

In more specific terms, and according to the original version 1.0.0 of the standard defined by Open Networking Foundation (2009), an OpenFlow-enabled Switch consists of a *flow table* and an OpenFlow *secure channel* to an external OpenFlow controller. Typically, the forwarding table maintains a list of flow entries. Each flow entry comprises *match fields* containing header values to match packets against, *counters* to update when packets match for flow statistics collection purposes and a set of *actions* to apply to matching packets.

Accordingly, all incoming packets processed by the switch are compared against the flow table, where flow entries match packets based on a priority order specified by the controller. In case a matching entry is found, the flow counter is incremented, and the actions associated with the specific flow entry are performed on the incoming packet belonging to that flow. According to the OpenFlow specification (Open Networking Foundation 2009), these actions may include forwarding a packet out on a specific port, dropping the packet, removing or updating packet headers, etc. If no match is found in the flow table, then the unmatched packet is encapsulated and sent over the secure channel to the controller, which decides how it should be processed. Among other possible actions, the controller may define a new flow for that packet by inserting new flow table entries.

1.2.2.2. *SDN control plane*

Regarded as the most fundamental building entity in the SDN architecture, the control plane consists of a centralized software controller that is responsible for handling communications between network applications and devices through open interfaces. More specifically, SDN controllers translate the requirements of the application layer down to the underlying data plane elements and give relevant information up to SDN applications.

The SDN control layer is commonly referred to as the network operating system (NOS) as it supports the network control logic and provides the application layer with an abstracted view of the global network, which contains enough information to specify policies while hiding all implementation details.

Typically, the control plane is logically centralized and yet implemented as a physically distributed system for scalability and reliability reasons, as discussed in sections 1.3 and 1.4. In a distributed SDN control configuration, east-westbound application programming interfaces (APIs) (Lin et al. 2015; Almadani et al. 2021) are required to enable multiple SDN controllers to communicate with each other and to exchange network information.

Despite the many attempts to standardize SDN protocols, there has been to date no standard for the east-west API, which remains proprietary for each controller vendor. Although a number of east-westbound communications occur only at the data store level and do not require additional protocol specifics, it is becoming increasingly advisable to standardize that communication interface in order to provide wider interoperability between different controller technologies in different autonomous SDN networks.

On the other hand, an east-westbound API standard requires advanced data distribution mechanisms and involves other special considerations. This brings about additional SDN challenges, some of which have been raised by the state-of-the art distributed controller platforms discussed in sections 1.3 and 1.4 but have yet to be fully addressed.

1.2.2.3. *SDN application plane*

The SDN application plane comprises SDN applications, which are control programs designed to implement the network control logic and strategies. This higher level plane interacts with the control plane via an open northbound API. In doing so, SDN applications communicate their network requirements to the SDN controller, which translates them into southbound-specific commands and forwarding rules dictating the behavior of the individual data plane devices. Routing, traffic engineering (TE), firewalls and load balancing are typical examples of common SDN applications running on top of existing controller platforms.

In the context of SDN, applications leverage the decoupling of the application logic from the network hardware along with the logical centralization of the network control to directly express the desired goals and policies in a centralized, high-level manner without being tied to the implementation and state-distribution details of the underlying networking infrastructure. Concurrently, SDN applications make use of the abstracted network view exposed through the northbound interface to consume the network services and functions provided by the control plane, according to their specific purposes.

That being said, the northbound API implemented by SDN controllers can be regarded as a network abstraction interface to applications, easing network programmability, simplifying control and management tasks and allowing for innovation. In contrast to the southbound API, the northbound API is not supported by an accepted standard.

Despite the broad variety of northbound APIs adopted by the SDN community (see Figure 1.2), we can classify them into two main categories.

– The first set involves simple and primitive APIs that are directly linked to the internal services of the controller platform. These implementations include:

- low-level ad hoc APIs that are proprietary and tightly dependent on the controller platform. Such APIs are not considered high-level abstractions as they allow developers to directly implement applications within the controller in a low-level manner. Deployed internally, these applications are tightly coupled with the controller and written in its native general-purpose language. NOX in C++ and POX in Python are typical examples of controllers that use their own basic sets of APIs;

- APIs based on Web services, such as the widely used REST API. This group of programming interfaces enables independent external applications (*clients*) to access the functions and services of the SDN controller (*server*). These applications can be written in any programming language and are not run inside the bundle hosting the controller software. Floodlight[1] is an example of an SDN controller that adopts an embedded northbound API based on REST.

– The second category contains higher level APIs that rely on domain-specific programming languages such as Frenetic (Foster et al. 2011; Kulkarni et al. 2021), Procera (Voellmy et al. 2012) and Pyretic (Monsanto et al. 2013; Kulkarni et al. 2021) as an indirect way for applications to interact with the controller. These APIs are designed to raise the level of abstraction in order to allow for the flexible development of applications and the specification of high-level network policies.

1.3. Physical classification of existing SDN control plane architectures

Despite the undeniable strengths of SDN, there have always been serious concerns about the ability to extend SDN to large-scale networks.

Some argue that these scalability limits are in effect linked to the protocol standards being used for the implementation of SDN. OpenFlow (McKeown et al. 2008; Costa et al. 2021) in particular, although recognized as a leading and widely deployed SDN southbound technology, is currently being rethought for potentially causing excessive overheads on switches (*switch bottleneck*). Scalable alternatives to the OpenFlow standard, which propose to revisit the delegation of control between the controller and the switches with the aim of reducing the reliance on SDN control plane, have been discussed in section 1.2.2.1.

Another entirely different approach to addressing the SDN scalability and reliability challenges – advocated here – is to physically distribute the SDN control plane. This has led to a first categorization of existing controller platforms into centralized and distributed architectures (see Figure 1.3). Please note that in Figures 1.3 and 1.4 controllers that present similar characteristics for the discussed comparison criteria are depicted in the same color.

1.3.1. *Physically centralized SDN control*

A physically centralized control plane consisting of a single controller for the entire network is a theoretically perfect design choice in terms of simplicity. However, a single controller system may not keep up with the growth of the network. It is likely to become overwhelmed (*controller bottleneck*) when dealing with an increasing number of requests while simultaneously struggling to achieve the same performance guarantees.

---

1 https://floodlight.atlassian.net/wiki/spaces/floodlightcontroller/overview.

A centralized SDN controller evidently does not meet the different requirements of large-scale, real-world network deployments. Data centers and service provider networks are typical examples of such large-scale networks, presenting different requirements in terms of scalability and reliability.

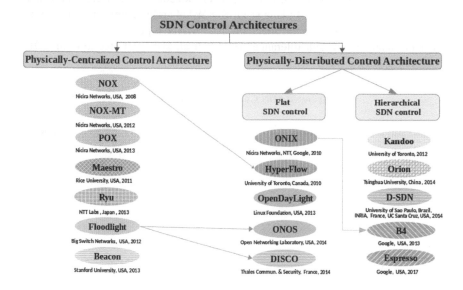

Figure 1.3. *Physical classification of SDN control plane architectures. For a color version of this figure, see www.iste.co.uk/ bannour/software.zip*

More specifically, a *data center network* involves tens of thousands of switching elements. Such a large number of forwarding elements, which can grow at a fast pace, is expected to generate a huge number of control events that are enough to overload a single centralized SDN controller (Yeganeh et al. 2013; Azodolmolky et al. 2013). Studies conducted in Benson et al. (2010) show important scalability implications (in terms of throughput) for centralized controller approaches. They demonstrate that multiple controllers should be used to scale the throughput of a centralized controller and meet the traffic characteristics within realistic data centers.

Unlike data centers, *service provider networks* are characterized by a modest number of network nodes. However, these nodes are usually geographically distributed, making the diameter of these networks very large (Yeganeh et al. 2013). This entails different types of controller scalability issues for centralized controller approaches, more specifically, high latencies. In addition to latency requirements, service provider networks have large numbers of flows that may generate overhead and bandwidth issues.

In general, wide area network (WAN) deployments typically impose strict resiliency requirements. In addition, they present higher propagation delays compared to data center networks. Obviously, a centralized controller design in an SD-WAN cannot achieve the desired failure resiliency and scale-out behaviors (Michel and Keller 2017; Sarmiento et al. 2021). Several studies have emphasized the need for a distributed control plane in an SD-WAN architecture. Indeed, they are focused on placing multiple controllers on real WAN topologies to benefit both control plane latency and fault tolerance (Heller et al. 2012; Ul Huque et al. 2017; Sminesh et al. 2019).

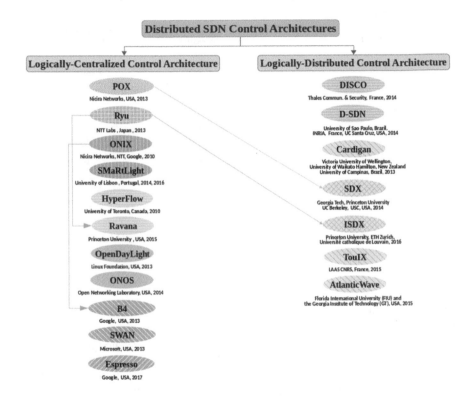

Figure 1.4. *Logical classification of distributed SDN control plane architectures. For a color version of this figure, see www.iste.co.uk/bannour/software.zip*

That said, the potential scalability, reliability and vulnerability concerns associated with centralized controller approaches have been further confirmed through studies (Shalimov et al. 2013; Karakus and Durresi 2017) on the behavior of state-of-the-art centralized SDN controllers such as NOX (Gude et al. 2008), Beacon (Erickson 2013) and Floodlight, within different networking environments.

In particular, NOX classic (Gude et al. 2008), the world's first-generation OpenFlow controller with an event-based programming model, is believed to be limited in terms of throughput. Indeed, it cannot handle a large number of flows, namely, a rate of 30k flow initiation events per second (Tavakoli et al. 2009; Karakus and Durresi 2017). Such a flow setup throughput may seem sufficient for an enterprise network, but it could be arguable for data-center deployments with high flow initiation rates (Benson et al. 2010). Improved versions of NOX have subsequently been developed by the same community (Nicira Networks), such as NOX-MT (Tootoonchian et al. 2012) for better performance and POX for a more developer-friendly environment.

However, while none of these centralized designs is believed to meet the above scalability and reliability requirements of large-scale networks, they have gained greater prominence as they were widely used for research and educational purposes.

Additionally, Floodlight, which is a very popular Java-based OpenFlow controller from Big Switch Networks, suffers from serious security and resiliency issues. For instance, Dhawan et al. (2015) reported that the centralized SDN controller is inherently susceptible to denial-of-service (DoS) attacks. Another subsequent version of Floodlight, called SE-Floodlight, has therefore been released to overcome these problems by integrating security applications. However, despite the introduced security enhancements aimed at shielding the centralized controller, the latter remains a potential weakness that compromises the whole network. In fact, the controller still maintains a single point of failure and bottlenecks even if its latest version is less vulnerable to malicious attacks.

On the other hand, given its obvious performance and functionality advantages, the open-source Floodlight has been extensively used to build other SDN controller platforms that support distributed architectures such as ONOS (Berde et al. 2014; Ruslan et al. 2020) and DISCO (Phemius et al. 2013).

1.3.2. Physically distributed SDN control

Alternatively, physically distributed control plane architectures have received increased research attention in recent years since they appeared as a potential solution to mitigate the issues brought about by centralized SDN architectures (poor scalability, single point of failure (SPOF), performance bottlenecks, etc.). As a result, various SDN control plane designs have been proposed in the recent literature. However, we discern two main categories of distributed SDN control architectures based on the physical organization of SDN controllers: a flat SDN control architecture and a hierarchical SDN control architecture (see Figure 1.3).

1.3.2.1. *Flat SDN control*

The flat structure implies the horizontal partitioning of the network into multiple areas, each of which is handled by a single controller in charge of managing a subset of SDN switches. There are several advantages to organizing controllers in a flat design, including reduced control latency and improved resiliency.

Onix (Koponen et al. 2010), HyperFlow (Tootoonchian and Ganjali 2010) and ONOS (Berde et al. 2014; Ruslan et al. 2020) are typical examples of flat, physically distributed controller platforms, initially designed to improve control plane *scalability* through the use of multiple interconnected controllers sharing a global network-wide view and allowing for the development of centralized control applications. However, each of these contributions takes a different approach to distribute controller states and provide control plane scalability.

Onix, for example, provides good scalability through additional partitioning and aggregation mechanisms. To be more specific, Onix partitions the NIB (Network Information Base), giving each controller instance responsibility for a subset of the NIB; it aggregates by making applications reduce the fidelity of information before sharing it between other Onix instances within the cluster. Similar to Onix, each ONOS instance (composing the cluster) that is responsible for a subset of network devices holds a portion of the network view that is also represented in a graph. In contrast to Onix and ONOS, every controller in HyperFlow has the global network view; thus, obtaining the illusion of control over the whole network. However, HyperFlow can be considered a scalable option for specific policies in which a small number of network events affect the global network state. In that case, scalability is ensured by propagating these (less frequent) selected events through the event propagation system.

Furthermore, different mechanisms are put in place by these distributed controller platforms to meet *fault tolerance* and *reliability* requirements in the event of failures or attacks.

Onix (Koponen et al. 2010) uses different recovery mechanisms depending on the detected failures. An Onix instance failure is predominantly handled by distributed coordination mechanisms among replicas, whereas network element/link failures are under the full responsibility of applications developed atop Onix. In addition, Onix is assumed to be reliable in terms of connectivity infrastructure failures because it can dedicate the failure recovery task to a separate management backbone that uses a multi-pathing protocol.

Similarly, HyperFlow (Tootoonchian and Ganjali 2010) focuses on ensuring resiliency and fault tolerance as a means of achieving availability. When a controller failure is discovered by the failure detection mechanisms deployed by its publish/subscribe WheelFS system (Stribling et al. 2009), HyperFlow reconfigures the affected switches and redirects them to another nearby controller instance (from a neighbor's site). Alongside this ability to tackle component failures, HyperFlow is resilient to network partitioning thanks to the partition tolerance property of WheelFS. In fact, in the presence of network partitioning, WheelFS partitions continue to operate independently, thus favoring availability.

Similarly, ONOS (Berde et al. 2014; Ruslan et al. 2020) considers fault tolerance to be a prerequisite for adopting SDN in service provider networks. ONOS's distributed control plane guards against controller instance failures by connecting, from the onset, each SDN switch to more than one SDN controller; its master controller and other backup controllers (from other domains) which may take over in the wake of master controller failures. Load balancing mechanisms are also provided to balance the mastership of switches among the controllers of the cluster for scalability purposes. In addition, ONOS incorporates additional recovery protocols, such as the Anti-Entropy protocol (Bianco et al. 2016), for healing from lost updates due to such controller crashes.

More recent SDN controller platform solutions (Chandrasekaran and Benson 2014; Shin et al. 2014; Katta et al. 2015; Yeganeh and Ganjali 2016; Chandrasekaran et al. 2016; Spalla et al. 2016; Mantas and Ramos 2019) focused specifically on improving fault tolerance in the distributed SDN control plane. Some of these works assumed a simplified flat design in which the SDN control was centralized. However, since the main focus was placed on the fault tolerance aspect, we believe that their ideas and their fault tolerance approaches can be leveraged in the context of medium- to large-scale SDNs in which the network control is physically distributed among multiple controllers.

In particular, Botelho et al. (2014) developed a hybrid SDN controller architecture that combines both passive and active replication approaches for achieving control plane fault tolerance. SMaRtLight adopts a simple Floodlight-based multi-controller design following OpenFlow 1.3, where one main controller (the primary) manages all network switches, and other controller replicas monitor the primary controller and serve as backups in case it fails.

This variant of a traditional passive replication system relies on an external data store that is implemented using a modern active replicated state machine (RSM) built with a Paxos-like protocol (BFT-SMaRt (Bessani et al. 2014)) to ensure fault tolerance and strong consistency. This shared data store is used for storing the network and application state (the common global NIB) and also for coordinating fault detection and leader election operations between controller replicas that run a lease management algorithm.

In the case of a failure of the primary controller, the elected backup controller starts reading the current state from the shared consistent data store in order to mitigate the cold-start (empty state) issue associated with traditional passive replication approaches, and thereby ensure a smoother transition to the new primary controller role.

The limited feasibility of the deployed controller fault tolerance strategy is warranted by the limited scope of the SMaRtLight solution, which is only intended for small- to medium-sized SDN networks. On the other hand, in large-scale deployments, adopting a simplified master–slave approach, and, more importantly, assuming a single main controller scheme in which one controller replica must retrieve all the network state from the shared data store in failure scenarios, have major disadvantages in terms of increased latency and failover time.

Similarly, the Ravana controller platform proposal (Katta et al. 2015) addresses the issue of recovering from complete fail-stop controller crashes. It offers the abstraction of a fault-free centralized SDN controller to unmodified control applications that are relieved of the burden of handling controller failures. Accordingly, network programmers write application programs for a single main controller, and the transparent master–slave Ravana protocol takes care of replicating the control logic – seamlessly and consistently – to other backup controllers for fault tolerance.

The Ravana approach deploys enhanced RSM techniques that are extended with switch-side mechanisms to ensure control messages are processed transactionally with ordered and exactly-once semantics, even in the presence of failures. The three Ravana prototype components, namely the Ryu-based controller (Ryu SDN Framework) runtime, the switch runtime, and the control channel interface, work cooperatively to guarantee the desired correctness and robustness properties of a fault-tolerant logically centralized SDN controller.

More specifically, when the master controller crashes, the Ravana protocol detects the failure within a short failover time and elects the standby slave controller to take over using Zookeeper-like (Hunt et al. 2010) failure detection and leader election mechanisms. The new leader finishes processing any logged events in order to catch up with the failed master controller state. Then, it registers with the affected switches in the role of the new master before proceeding with normal controller operations.

1.3.2.2. *Hierarchical SDN control*

The hierarchical SDN control architecture assumes that the network control plane is vertically partitioned into multiple levels (layers) depending on the required services. According to Liu et al. (2015) and Amiri et al. (2019), a hierarchical organization of the control plane can improve SDN scalability and performance.

To improve *scalability*, Hassas Yeganeh and Ganjali (2012) assumes a hierarchical two-layer control structure that partitions control applications into local and global. Contrary to DevoFlow (Curtis et al. 2011; Abuarqoub 2020) and DIFANE (Yu et al. 2010; Abuarqoub 2020), Kandoo proposes to reduce the overall stress on the control plane without the need to modify OpenFlow switches. Instead, it establishes a two-level hierarchical control plane, where frequent events occurring near the data path are handled by the bottom layer (local controllers with no interconnection running local applications) and non-local events requiring a network-wide view are handled by the top layer (a logically centralized root controller running non-local applications and managing local controllers).

Despite the obvious scalability advantages of such a control plane configuration, in which local controllers can scale linearly as they do not share information, Kandoo did not envisage *fault tolerance* and resiliency strategies to protect itself from potential failures and attacks in the data and control planes. In addition, from a developer perspective, Kandoo imposes some Kandoo-specific conditions on the control applications developed on top of it in such a way that makes them aware of its existence.

On the other hand, Google's B4 (Jain et al. 2013; Kumar et al. 2015), a private intra-domain software-defined WAN connecting their data centers across the planet, proposes a two-level hierarchical control framework for improving *scalability*. At the lower layer, each data center site is handled by an Onix-based (Koponen et al. 2010) SDN controller hosting local site-level control applications. These site controllers are managed by a global *SDN Gateway* that collects network information from multiple sites through site-level TE services and sends them to a logically centralized *TE server*, which also operates at the upper layer of the control hierarchy. Based on an abstract topology, the latter enforces high-level TE policies that are mainly aimed at optimizing bandwidth allocation between competing applications across the different data center sites. That being said, the TE server programs these forwarding rules at the different sites through the same gateway API. These TE entries will be installed in higher priority switch forwarding tables alongside the standard shortest-path forwarding tables. In this context, it is worth mentioning that the *topology abstraction,* which consists of abstracting each site into a *super-node* with an aggregated *super-trunk* to a remote site, is key to improving the scalability of the B4 network. Indeed, this abstraction hides the details and complexity from the logically centralized TE controller, thereby allowing it to run protocols at a coarse granularity based on a global controller view and, more importantly, preventing it from becoming a serious performance bottleneck.

Unlike Kandoo (Hassas Yeganeh and Ganjali 2012), B4 (Jain et al. 2013) deploys robust *reliability* and *fault tolerance* mechanisms at both levels of the control hierarchy in order to enhance the B4 system availability. These mechanisms have been specially enhanced after experiencing a large-scale B4 outage. In particular, Paxos (Chandra et al. 2007) is used for detecting and handling the primary controller failure within each data center site by electing a new leader controller among a set of reachable standby instances. On the other hand, network failures at the upper layer are addressed by the logically centralized TE controller, which adapts to failed or unresponsive site controllers in the bandwidth allocation process. Additionally, B4 is resilient against other failure scenarios in which the upper level TE controller encounters major problems in reaching the lower level site controllers (e.g. TE operation/session failures). Moreover, B4 guards against the failure of the logically centralized TE controller by geographically replicating TE servers across multiple WAN sites (one master TE server and four secondary hot standbys). Finally, another fault recovery mechanism is used in case the TE controller service itself faces serious problems. This mechanism stops the TE service and enables the standard shortest-path routing mechanism to be an independent service.

Similarly, Espresso (Yap et al. 2017) is another interesting SDN contribution which represents the latest and more challenging pillar of Google's SDN strategy. Building on the previous three layers of this strategy (the *B4* WAN (Jain et al. 2013), the *Andromeda* NFV stack and the *Jupiter* data center interconnect), *Espresso* extends the SDN approach to the peering edge of Google's network, where it connects to other networks worldwide. Considered a large-scale SDN deployment for the public Internet, Espresso, which has been in production for more than 2 years, routes over 22% of Google's total traffic to the Internet. More specifically, the Espresso technology allows Google to dynamically choose from where to serve content for individual users based on real-time measurements of end-to-end network connections.

To deliver unprecedented *scale-out* and efficiency, Espresso assumes a hierarchical control plane design split between *global controllers* and *local controllers*, which perform different functions. In addition, Espresso's software programmability design principle externalizes features into software, thereby exploiting commodity servers for scale.

Moreover, Espresso achieves higher availability (*reliability*) when compared to existing router-centric Internet protocols. It supports a fail static system in which the local data plane maintains the last known good state to allow for control plane unavailability, without impacting data plane and BGP peering operations. Finally, another important feature of Espresso is that it provides full *interoperability* with the rest of the Internet and the traditional heterogeneous peers.

1.4. Logical classification of existing SDN control plane architectures

Aside from the physical classification, we can categorize distributed SDN control architectures according to the way knowledge is disseminated among controller instances (the *consistency* challenge) into logically centralized and logically distributed architectures (see Figure 1.4). This classification has been recently adopted by Benamrane et al. (2015).

1.4.1. *Logically centralized SDN control*

1.4.1.1. *Onix and SMaRtLight*

Both Onix (Koponen et al. 2010) and SMaRtLight (Botelho et al. 2014) are logically centralized controller platforms that achieve controller state redundancy through state replication. However, the main difference is that Onix uses a distributed data store, while SMartLight uses a centralized data store to replicate the shared network state. They also deploy different techniques for sharing knowledge and maintaining a consistent network state.

Onix is a distributed control platform for large-scale production networks which stands out from previous proposals by providing a simple general-purpose API, a central NIB abstraction and standard state distribution primitives for easing the implementation of network applications.

In more specific terms, Onix uses the NIB data structure to store the global network state (in the form of a network graph) that is distributed across running Onix instances and synchronized through Onix's built-in state distribution tools, according to different levels of consistency as dictated by application requirements. In fact, in addition to interacting with the NIB at runtime, network applications on top of Onix initially configure their own data storage and dissemination mechanisms by choosing between two data store options already implemented by Onix in the NIB. A replicated transactional database that guarantees strong consistency at the cost of good performance for persistent but slowly changing data (state), and a high-performance memory-only distributed hash table (DHT) for volatile data that does not require strict consistency.

While the main advantage of Onix is its programmatic framework – created for the flexible development of control applications with desired trade-offs between performance and state consistency (strong/eventual) – it carries the limitations of eventually consistent systems that rely on application-specific logic to detect network state inconsistencies for the eventually consistent data and provide conflict resolution methods for handling them.

As mentioned in section 1.3.2.1, SMaRtLight is a fault-tolerant logically centralized master–slave SDN controller platform in which a single controller is in charge of all network decisions. This main controller is supported by backup controller replicas that should have a synchronized network view in order to take over the network control in case of the primary failure. All controller replicas are coordinated through a shared data store that is kept fault-tolerant and strongly consistent using an implementation of RSM.

Consistency between the master and backup controllers is guaranteed by replicating each change in the network image (NIB) of the master in the shared data store before modifying the state of the network. However, such synchronization updates generate additional time overheads and have a drastic impact on the controller's performance. To address this issue, the controllers keep a local cache (maintained by one active primary controller at any time) to avoid accessing the shared data store for read operations. By keeping the local cache and the data store consistent even in the presence of controller failures, the authors claim that their simple master–slave structure, in the context of small to medium-sized networks, achieves a balance between consistency and fault tolerance while keeping performance at an acceptable level.

1.4.1.2. *HyperFlow and Ravana*

Both HyperFlow (Tootoonchian and Ganjali 2010) and Ravana (Katta et al. 2015) are logically centralized controller platforms that achieve state redundancy through event replication. Despite their similarities in building the application state, one difference is that the Ravana protocol is completely transparent to control applications, while HyperFlow requires minor modifications to applications. In addition, while HyperFlow is eventually consistent favoring availability, Ravana ensures strong consistency guarantees.

More specifically, HyperFlow (Tootoonchian and Ganjali 2010) is an extension of NOX into a distributed event-based control plane where each NOX-based controller manages a subset of OpenFlow network switches. It uses an event-propagation publish/subscribe mechanism based on the distributed WheelFS (Stribling et al. 2009) file system for propagating selected network events and maintaining the global network-wide view across controllers. Accordingly, the HyperFlow controller application instance running on top of an individual NOX controller selectively publishes relevant events that affect the network state and receives events on subscribed channels to other controllers. Then, other controllers locally replay all the published events in order to reconstruct the state and achieve the synchronization of the global view.

By this means, all controller instances make decisions locally and individually (without contacting remote controller instances). Indeed, they operate based on their synchronized eventually consistent network-wide view as if they are in control of the entire network. Through this synchronization scheme, HyperFlow achieves the goal of minimizing flow setup times and also congestion; in other words, cross-site traffic required to synchronize the state among controllers. However, the potential downside of HyperFlow is related to the performance of the publish/subscribe system, which can only deal with non-frequent events. In addition, HyperFlow does not guarantee a strict ordering of events and does not handle consistency problems. This makes the scope of HyperFlow restricted to applications that do not require a strict event ordering with strict consistency guarantees.

To correctly ensure the abstraction of a "logically centralized SDN controller", an elaborate fault-tolerant controller platform called Ravana (Katta et al. 2015) extended beyond the requirements for controller state consistency to include that for switch state consistency under controller failures.

Maintaining such strong levels of consistency in both controllers and switches in the presence of failures requires the entire event-processing cycle to be handled as a transaction in accordance with the following properties: (i) events are processed in the same total order at all controller replicas so that controller application instances reach the same internal state; (ii) events are processed exactly once across all the controller replicas; and (iii) commands are executed exactly once on the switches.

To achieve such design goals, Ravana follows a Replicated State Machine (RSM) approach but extends its scope to deal with switch state consistency under failures. Indeed, while Ravana permits unmodified applications to run in a transparent fault-tolerant environment, it requires modifications to the OpenFlow protocol, and it makes changes to current switches instead of involving them in a complex consensus protocol.

To be more specific, Ravana uses a two-stage replication protocol that separates the reliable logging of the master's event delivery information (stage 1) from the logging of the master's event-processing transaction completion information (stage 2) in the shared in-memory log (using Viewstamped Replication (Oki and Liskov 1988)) in order to guarantee consistency under joint switch and controller failures. In addition, it adds explicit acknowledgment messages to the OpenFlow 1.3 protocol and implements buffers on existing switches for event retransmission and command filtering. The main objective of the addition of these extensions and mechanisms is to guarantee the exactly-once execution of any event transaction on the switches during controller failures.

Such strong correctness guarantees for a logically centralized controller under Ravana come at the cost of generating additional throughput and latency overheads, which can be reduced to a reasonable extent with specific performance optimizations. Since the Ravana runtime is completely transparent and oblivious to control applications, achieving relaxed consistency requirements for the sake of improved availability, as required by some specific applications, involves consideration of new mechanisms that relax some of the correctness constraints on Ravana's design goals.

A similar approach to Ravana (Katta et al. 2015) was adopted by Mantas and Ramos (2016) to achieve a consistent and fault-tolerant SDN controller. In their work, the authors claim to retain the same requirements expressed by Ravana, namely, the transparency, reliability, consistency and performance guarantees but without requiring changes to the OpenFlow protocol or to existing switches.

Similarly, Kandoo (Hassas Yeganeh and Ganjali 2012) falls into the category of logically centralized controllers that distribute the control state by propagating network events. Indeed, at the top layer of its hierarchical design, Kandoo assumes a logically centralized root controller for handling global and rare network events. Since its aim was to preserve scalability without changing the OpenFlow devices, Kandoo did not focus on knowledge distribution mechanisms for achieving network state consistency.

1.4.1.3. *ONOS and OpenDayLight*

ONOS[2] and OpenDayLight (ODL)[3] (Bondkovskii et al. 2016; Badotra and Panda 2020) represent another category of logically centralized SDN solutions, which set themselves apart from state-of-the-art distributed SDN controller platforms by offering community-driven open-source frameworks and providing the full functionalities of network operating systems. Despite their obvious similarities, these prominent Java-based projects present major differences in terms of structure, target customers, focus areas and inspirations.

Unlike ODL (Badotra and Panda 2020), which is applicable to different domains, ONOS (Berde et al. 2014; Ruslan et al. 2020) from ON.LAB is specifically targeted toward service providers and is thus architected to meet their carrier-grade requirements in terms of scalability, high-availability and performance. In addition to the high-level northbound abstraction (a global network view and an application intent framework) and the pluggable southbound abstraction (supporting multiple protocols), ONOS, in the same way as Onix and HyperFlow, offers state dissemination mechanisms (Muqaddas et al. 2016) to achieve a consistent network state across the distributed cluster of ONOS controllers, a required or highly desirable condition for network applications to run correctly.

More specifically, ONOS's distributed core eases the state management and cluster coordination tasks for application developers by providing them with an available set of core building blocks for dealing with different types of distributed control plane state, including a *ConsistentMap* primitive for state requiring strong consistency and an *EventuallyConsistentMap* for state tolerating weak consistency.

In particular, applications that favor performance over consistency store their state in the shared eventually consistent data structure that uses optimistic replication assisted by the gossip-based anti-entropy protocol (Bianco et al. 2016). For example, the global network topology state, which should be accessible to applications with minimal delays, is managed by the *network topology store* according to this eventual consistency model. Recent releases of ONOS treat the network topology view as an in-memory state machine graph. The latter is built and updated in each SDN

---

2 https://opennetworking.org/onos/.
3 https://opendaylight.org/.

controller by applying local topology events and replicating them to other controller instances in the cluster in an order-aware fashion based on the events' logical timestamps. Potential conflicts and loss of updates due to failure scenarios are resolved by the anti-entropy approach (Bianco et al. 2016), in which each controller periodically compares its topology view with that of another randomly selected controller in order to reconcile possible differences and recover from stale information.

On the other hand, state imposing strong consistency guarantees is managed by the second data structure primitive built using RAFT (Ongaro and Ousterhout 2014), a protocol that achieves consensus via an elected leader controller in charge of replicating the received log updates to follower controllers and then committing these updates upon receipt of confirmation from the majority. The mapping between controllers and switches, which is handled by ONOS's *mastership store,* is an example of a network state that is maintained in a strongly consistent manner.

Administered by the Linux Foundation and backed by the industry, ODL is a generic and general-purpose controller framework that, unlike ONOS, was conceived to accommodate a wide variety of applications and use cases concerning different domains (e.g. Data Center, Service Provider and Enterprise). One important architectural feature of ODL is its YANG-based Model-Driven Service Abstraction Layer (MD-SAL) which allows for easy and flexible incorporation of network services requested by the higher layers via the northbound interface (OSGi framework and the bidirectional RESTful Interfaces) irrespective of the multiple southbound protocols used between the controller and the heterogeneous network devices.

The main focus of ODL was to accelerate the integration of SDN in legacy network environments by automating the configuration of traditional network devices and enabling their communication with OpenFlow devices. As a result, the project was perceived as adopting vendor-driven solutions that mainly aim to preserve the brands of legacy hardware. This represents a broad divergence from ONOS, which envisages a carrier-grade SDN platform with enhanced performance capabilities to explore the full potential of SDN and demonstrate its real value.

The latest releases of ODL provided a distributed SDN controller architecture referred to as ODL clustering. In contrast to ONOS, ODL did not offer various consistency models for different types of network data. All the data shared across the distributed cluster of ODL controllers for maintaining the logically centralized network view are handled in a strongly consistent manner using the RAFT consensus algorithm (Ongaro and Ousterhout 2014) and the Akka framework.

1.4.1.4. *B4 and SWAN*

Google's B4 (Jain et al. 2013) network leverages the logical centralization enabled by the SDN paradigm to deploy centralized TE in coexistence with the standard shortest-path routing for the purpose of increasing the utilization of the inter-data center links (near 100%) compared to conventional networks and thereby enhancing network efficiency and performance. As previously explained in section 1.3.2.2, the logically centralized *TE server* uses the network information collected by the centralized *SDN gateway* to control and coordinate the behavior of site-level SDN controllers based on an abstracted topology view. The main task of the TE server is indeed to optimize the allocation of bandwidth among competing applications (based on their priority) across the geographically distributed data center sites.

That being said, we implicitly assume the presence of a specific consistency model used by the centralized SDN gateway for handling the distributed network state across the data center site controllers and ensuring the centralized TE application runs correctly based on a consistent network-wide view. However, very little information has been provided on the level of consistency adopted by the B4 system. In fact, one potential downside of the SDN approach followed by Google could be that it is too customized and tailored to their specific network requirements, given no general control model has been proposed for future use by other SDN projects.

Similarly, Microsoft has presented SWAN (Hong et al. 2013) as an intra-domain software-driven WAN deployment that takes advantage of the logically centralized SDN control using a global TE solution to significantly improve the efficiency, reliability and fairness of their inter-DC WAN. Like Google, Microsoft did not provide much information about the control plane state consistency updates.

1.4.2. *Logically distributed SDN control*

The potential of the SDN paradigm has been fully explored within single administrative domains like data centers, enterprise networks, campus networks and even WANs, as discussed in section 1.4.1. Indeed, the main pillars of SDN – the decoupling between the control and data planes together with the consequent ability to program the network in a logically centralized manner – have unleashed productive innovation and novel capabilities in the management of such intra-domain networks. These benefits include the effective deployment of new domain-specific services as well as the improvement of standard control functions, following the SDN principles such as intra-domain routing and TE. RCP (Caesar et al. 2005) and RouteFlow (Rothenberg et al. 2012) are practical examples of successful intra-AS platforms that use OpenFlow to provide conventional IP routing services in a centralized manner.

However, this main feature of logically centralized control, which has been leveraged by most SDN solutions to improve network management at the intra-domain level, cannot be fully exploited for controlling heterogeneous networks involving multiple autonomous systems (ASes) under different administrative authorities (e.g. the Internet). In this context, recent works have considered extending the SDN scheme to such inter-domain networks while remaining compatible with their distributed architecture. In this section, we shed light on these SDN solutions that adopted a logically distributed architecture in accordance with legacy networks. For this reason, we place them in the category of logically distributed SDN platforms as opposed to the logically centralized ones mainly used for intra-domain scenarios.

1.4.2.1. *DISCO and D-SDN*

The DISCO project (Phemius et al. 2013), for example, suggests a logically distributed control plane architecture that operates in such multi-domain heterogeneous environments, more precisely WANs and overlay networks. Built on top of Floodlight, each DISCO controller administers its own SDN network domain and interacts with other controllers to provide end-to-end network services. This inter-AS communication is ensured by a unique lightweight control channel to share summary network-wide information.

The most obvious contribution of DISCO lies in the separation between intra-domain and inter-domain features of the control plane, while each type of feature is performed by a separate part of the DISCO architecture. The intra-domain modules are responsible for ensuring the main functions of the controller such as monitoring the network and reacting to network issues, and the inter-domain modules (messenger, agents) are designed to enable message-oriented communication between neighbor domain controllers. Indeed, the AMQP-based messenger offers a distributed publish/subscribe communication channel used by agents, which operates at the inter-domain level by exchanging aggregated information with intra-domain modules. DISCO was assessed on an emulated environment according to three use cases: inter-domain topology disruption, end-to-end service priority request and virtual machine migration.

The main advantage of the DISCO solution is the potential to adapt it to large-scale networks with different ASes such as the Internet (Benamrane et al. 2015). However, we believe that there are also several drawbacks associated with such a solution including the static non-evolving decomposition of the network into several independent entities, which is in contrast to emerging theories such as David D. Clark's theory (Clark et al. 2003) about the network being manageable by an additional high-level entity known as the Knowledge Plane. Moreover, following the DISCO architecture, network performance optimization becomes a local task dedicated to local entities with different policies, each of which acts in its own best interest at the expense of the general interest. This leads to local optima rather than the global optimum which achieves the global network performance. Additionally, from the DISCO perspective, one SDN controller is responsible for one independent domain. However, an AS is usually too large to be handled by a single controller. Finally, DISCO did not provide appropriate reliability strategies suited to its geographically distributed architecture. In fact, in the event of a controller failure, it could be inferred that a remote controller instance will be in charge of the subset of affected switches, thereby resulting in a significant increase in the control plane latency. In our opinion, a better reliability strategy would involve local per-domain redundancy. Local controller replicas should indeed take over and serve as backups in case the local primary controller fails.

Similarly, INRIA's D-SDN (Santos et al. 2014) enables a logical distribution of the SDN control plane based on a hierarchy of *Main Controllers* and *Secondary Controllers*, matching the organizational and administrative structure of the current and future Internet. In addition to dealing with levels of control hierarchy, another advantage of D-SDN over DISCO is related to its enhanced security and fault tolerance features.

1.4.2.2. *SDX-based controllers*

Unlike DISCO, which proposes per-domain SDN controllers with inter-domain functions to allow autonomous end-to-end flow management across SDN domains, recent trends have considered deploying SDN at Internet eXchange Points (IXPs), thus giving rise to the concept of Software-Defined eXchanges (SDXes). These SDXes are used to interconnect participants of different domains via a shared software-based platform. This platform is usually aimed at bringing innovation to traditional peering, easing the implementation of customized peering policies and enhancing the control over inter-domain traffic management.

Prominent projects adopting this vision of software-defined IXPs and implementing it in real production networks include Google's Cardigan in New Zealand (Stringer et al. 2014), SDX at Princeton (Gupta et al. 2014), CNRS's French TouIX (Lapeyrade et al. 2016) (the European ENDEAVOUR project) and the AtlanticWave-SDX (Heidi Morgan 2015). Here, we chose to focus on the SDX project at Princeton since we believe in its potential for demonstrating the capabilities of SDN to innovate IXPs and for bringing answers to deploying SDX in practice.

The SDX project (Gupta et al. 2014; Mostafaei et al. 2021) takes advantage of SDN-enabled IXPs to fundamentally improve wide-area traffic delivery and enhance conventional inter-domain routing protocols that lack the required flexibility for achieving various TE tasks. Today's BGP is indeed limited to destination-based routing, it has a local forwarding influence restricted to immediate neighbors and it deploys indirect mechanisms for controlling path selection. To overcome these limitations, SDX relies on SDN features to ensure fine-grained, flexible and direct expression of inter-domain control policies, thereby enabling a wider range of valuable end-to-end services such as inbound TE, application-specific peering, server load balancing and traffic redirection through middle-boxes.

The SDX architecture consists of a smart SDX controller handling both SDX policies (*policy compiler*) and BGP routes (*route server*), conventional Edge routers and an OpenFlow-enabled switching fabric. The main idea behind this implementation is to allow participant ASes to compose their own policies in a high-level (using Pyretic) and independent manner (through the virtual switch abstraction), and then send them to the SDX controller. The latter is in charge of compiling these policies to SDN forwarding rules while taking into account BGP information.

In addition to offering this high-level softwarized framework, which is easily integrated into the existing infrastructure while maintaining good interoperability with its routing protocol, SDX also stands out from similar solutions like Cardigan (Stringer et al. 2014) because of the efficient mechanisms used for optimizing control and data plane operations. In particular, the scalability challenges faced by SDX under realistic scenarios have been further investigated by iSDX (Gupta et al. 2016; Abdullahi et al. 2021), an enhanced Ryu-based version of SDX intended to operate at the scale of large industrial IXPs.

However, one major drawback of the SDX contribution is that it is limited to the participant ASes being connected via the software-based IXP, implying that non-peering ASes would not benefit from the routing opportunities offered by SDX. Moreover, while solutions built on SDX use central TE policies for augmenting BGP and promote a logical centralization of the routing control plane at the IXP level, SDX controllers are still logically decentralized at the inter-domain level since no information is shared between them about their respective interconnected ASes. This brings us back to the same problem we pointed out for DISCO (Phemius et al. 2013) regarding end-to-end traffic optimization being a local task for each part of the network.

To remedy this issue, some recent works (Kotronis et al. 2015) have considered centralizing the whole inter-domain routing control plane to improve BGP convergence by outsourcing the control logic to a multi-AS routing controller with a "bird's-eye view" over multiple ASes.

It is also worth mentioning that SDX-based controllers face several limitations in terms of both security and reliability.

Because the SDX controller is the central element in the SDX architecture, security strategies must focus on securing the SDX infrastructure by protecting the SDX controller against cyber attacks and by authenticating any access to it. In particular, Chung et al. (2015) argue that SDX-based controllers are subjected to the potential vulnerabilities introduced by SDN in addition to the common vulnerabilities associated with classical protocols. In this respect, they distinguish three types of current SDX architectures and discuss the security concerns involved.

In their opinion, Layer-3 SDX (Stringer et al. 2014; Gupta et al. 2014) will inherit all BGP vulnerabilities, Layer-2 SDX (Internet2) will obtain the vulnerabilities of a shared Ethernet network and SDN SDX (Lin et al. 2015) will also bring controller vulnerabilities like DDoS attacks, comprised controllers and malicious controller applications. Moreover, the same studies (Chung et al. 2015) point out that SDX-based controllers require security considerations with respect to policy isolation between different SDX participants.

Finally, since the SDX controller becomes a potential single point of failure, fault tolerance and resiliency measures should be taken into account when building an SDX architecture. While the distributed peer-to-peer SDN SDX architecture (Chung et al. 2016) is inherently resilient, centralized SDX approaches should incorporate fault tolerance mechanisms such as those discussed in section 2.3 and should also leverage the existing fault tolerant distributed SDN controller platforms (Berde et al. 2014; Ruslan et al. 2020; Das et al. 2020).

1.5. Conclusion

In this chapter, we have provided a detailed analysis of state-of-the-art distributed SDN controller platforms. We assessed their architecture components and design patterns, and classified them in novel ways (physical and logical classifications) in order to provide useful guidelines for SDN research and deployment initiatives.

Additionally, our extensive analysis of these SDN platform proposals has enabled us to gain a thorough understanding of their advantages and drawbacks and, most importantly, to develop a critical awareness of the challenges facing distributed control in SDNs. These open challenges are further discussed in Chapter 2.

2

Decentralized SDN Control: Major Open Challenges

2.1. Introduction

While offering the potential to transform and improve current networks, the SDN initiative is still in the early stages of addressing the wide range of challenges involving different disciplines. In particular, distributed control of SDNs faces a series of pressing challenges that require our special consideration.

This chapter provides a thorough discussion of the major challenges of distributed SDN control, along with some insights into emerging and future trends in this area. These challenges include the issues of scalability (section 2.2), reliability (section 2.3), state consistency (section 2.4), interoperability (section 2.5), monitoring and security (section 2.6).

In the previous chapter, we surveyed the most prominent state-of-the-art distributed SDN controller platforms and, more importantly, we discussed the different approaches adopted to tackle the above challenges and proposed potential solutions. Table 2.1 gives a brief summary of the main features and key performance indicators (KPIs) of the discussed SDN controllers. Physically centralized controllers such as NOX (Gude et al. 2008), POX and FloodLight suffer from scalability and reliability issues. Solutions such as DevoFlow (Curtis et al. 2011; Abuarqoub 2020) and DIFANE (Yu et al. 2010; Abuarqoub 2020) attempted to solve these scalability issues by rethinking the OpenFlow protocol, whereas most SDN groups geared their focus toward distributing the control plane. While some of the distributed SDN proposals, such as Hassas Yeganeh and Ganjali (2012), promoted a hierarchical organization of the control plane to further improve scalability, other alternatives opted for a flat organization for increased reliability and performance (latency). On the other hand, distributed platforms such as Onix (Koponen et al. 2010), HyperFlow

(Koponen et al. 2010), ONOS (Berde et al. 2014; Ruslan et al. 2020) and OpenDaylight (Badotra and Panda 2020) focused on building consistency models for their logically centralized control plane designs. In particular, Onix (Koponen et al. 2010) chose DHT and transactional databases for network state distribution over the Publish/Subscribe system used by HyperFlow (Tootoonchian and Ganjali 2010). Another different class of solutions has been recently introduced by DISCO, which promoted a logically distributed control plane based on existing ASs within the Internet.

In Chapter 1, we classified these existing controllers according to the physical organization of the control plane (*physical classification*) and, alternatively, according to the way knowledge is shared in distributed control plane designs (*logical classification*). Furthermore, within each of these classifications, we performed another internal classification based on the similarities between competing SDN controllers (the *color classification* shown in Figures 1.3 and 1.4).

As mentioned above, it is obvious that there are various approaches to building a distributed SDN architecture; some of these approaches met certain performance criteria better than others but failed in some other aspects. It is clear that none of the proposed SDN controller platforms met all the discussed challenges and fulfilled all the KPIs required for an optimal deployment of SDN. At this stage, and building on these previous efforts, we communicate our vision of a distributed SDN control model by going through some of the major open challenges (see Figure 2.1), identifying the best ways of solving them, and envisaging future opportunities.

2.2. Scalability

Scalability concerns in SDN may arise from the decoupling between the control and data planes (Jammal et al. 2014; Rana et al. 2019) and the centralization of the control logic in a software-based controller. In fact, as the network grows in size (e.g. switches, hosts, and so on), the centralized SDN controller becomes highly solicited (in terms of events/requests) and thus overloaded (in terms of bandwidth, processing power and memory). Furthermore, when the network scales up in terms of both size and diameter, communication delays between the SDN controller and the network switches may become high, thus affecting flow-setup latencies. This may also cause congestion in both the control and data planes and may generate longer failover times (Karakus and Durresi 2017).

However, since control plane scalability in SDN is commonly assessed in terms of both *throughput* (the number of flow requests handled per second) and *flow setup latency* (the delay in responding to flow requests) metrics (Karakus and Durresi 2017), a single physically centralized SDN controller may not specifically fulfill the performance requirements (with respect to these metrics) of large-scale networks compared to small- or medium-scale networks (see section 1.3.1).

	Control plane architecture	Control plane design	Programming language	Scalability	Reliability	Consistency
NOX (Gude et al. 2008)	Physically Centralized	–	C++	Very limited	Limited	Strong
POX	Physically Centralized	–	Python	Very limited	Limited	Strong
Floodlight	Physically Centralized	–	Java	Very limited	Limited	Strong
SMaRtLight (Botelho et al. 2014)	Physically Centralized	–	Java	Very limited	Very good	Strong
Ravana (Katta et al. 2015)	Physically Centralized	–	Python	Limited	Very good	Strong
ONIX (Koponen et al. 2010)	Physically Distributed Logically Centralized	Flat	Python C	Very good	Good	Weak Strong
HyperFlow (Tootoonchian and Ganjali 2010)	Physically Distributed Logically Centralized	Flat	C++	Good	Good	Eventual
ONOS (Berde et al. 2014)	Physically Distributed Logically Centralized	Flat	Java	Very good	Good	Weak Strong
OpenDayLight	Physically Distributed Logically Centralized	Flat	Java	Very good	Good	Strong
B4 (Jain et al. 2013)	Physically Distributed Logically Centralized	Hierarchical	Python C	Good	Good	N/A
Kandoo (Hassas Yeganeh and Ganjali 2012)	Physically Distributed Logically Centralized	Hierarchical	C C++ Python	Very good	Limited	N/A
DISCO (Phemius et al. 2013)	Physically Distributed Logically Distributed	Flat	Java	Good	Limited	Strong (inter-domain)
SDX (Gupta et al. 2014)	Physically Distributed Logically Distributed	Flat	Python	Limited	N/A	Strong
DevoFlow (Curtis et al. 2011)	Physically Distributed Logically Centralized	N/A	Java	Good	N/A	N/A
DIFANE (Yu et al. 2010)	Physically Distributed Logically Centralized	N/A	–	Good	N/A	N/A

Table 2.1. *Main characteristics of the discussed SDN controllers*

Figure 2.1. *The main challenges of distributed SDN control. For a color version of this figure, see www.iste.co.uk/bannour/software.zip*

2.2.1. *Data plane extensions*

One way to alleviate some of these scalability concerns is to extend the responsibilities of the data plane in order to relieve the load on the SDN controller (Hohlfeld et al. 2018). The main drawback of this method is that it imposes some modifications to the design of OpenFlow switches. Indeed, despite the advantages linked to the flexibility and innovation brought to network management, OpenFlow (McKeown et al. 2008) suffers from scalability and performance issues that stem mainly from pushing all network intelligence and control logic to the centralized OpenFlow controller, thus restricting the task of OpenFlow switches to a dumb execution of forwarding actions.

To circumvent these limitations, several approaches (Yu et al. 2010; Curtis et al. 2011; Bianchi et al. 2014; Bosshart et al. 2014; Abuarqoub 2020) suggest revisiting the delegation of control between the controller and switches and introducing new SDN switch southbound interfaces.

Notably, DevoFlow (Curtis et al. 2011; Abuarqoub 2020) claims to minimize switch-to-controller interactions by introducing new control mechanisms inside switches. In this way, switches can make local control decisions when handling frequent events, without involving controllers whose primary tasks will be limited to keeping centralized control over far fewer significant events that require network-wide visibility. Despite introducing innovative ideas, the DevoFlow alternative has been mainly criticized for imposing major modifications to switch designs (Hassas Yeganeh and Ganjali 2012).

On the other hand, *stateful* approaches (Moshref et al. 2014; Bianchi et al. 2014; Arashloo et al. 2016; Bifulco and Rétvári 2018; Zhang et al. 2021), as opposed to the original *stateless* OpenFlow abstraction, motivate the need to delegate some stateful control functions back to switches in order to offload the SDN controller. These approaches face the challenging dilemma of programming stateful devices (evolving the data plane) while retaining the simplicity, generality and vendor-agnostic features offered by the OpenFlow abstraction. In particular, the OpenState proposal (Bianchi et al. 2014) is a stateful platform-independent data plane extension of the current OpenFlow match/action abstraction supporting a finite-state machine (FSM) programming model called Mealy Machine, in addition to the flow programming model adopted by OpenFlow. This model is implemented inside the OpenFlow switches using additional *state tables* in order to reduce the reliance on remote controllers for applications involving local states like MAC learning operations and port-knocking on a firewall.

Despite having the advantage of building on the adaptation activity of the OpenFlow standard and leveraging its evolution using the (stateful) extensions provided by recent versions (version 1.3 and 1.4), OpenState faces significant challenges regarding the implementation of a stateful extension for programming the forwarding behavior inside switches, while following an OpenFlow-like implementation approach. The feasibility of the hardware implementation of OpenState has been addressed in Pontarelli et al. (2015). Finally, the same authors extended their work into a more general and expressive abstraction of OpenState called OPP (Bianchi et al. 2016). This supports a full extended finite-state machine (XFSM) model, thereby enabling a broader range of applications and complex stateful flow processing operations.

In a similar way, the approach presented in Bifulco et al. (2016) and Michel et al. (2021) explored delegating some parts of the controller functions involving packet generation tasks to OpenFlow switches in order to address both switch and controller scalability issues. The InSP API was introduced as a generic API that extends OpenFlow to allow for the programming of autonomous packet generation operations inside switches such as ARP and ICMP handling. The proposed OpenFlow-like abstractions include an *InSP Instruction* for specifying the actions the switch should apply to a packet generated after a triggering event and a *Packet Template Table (PTE)* for storing the content of any packet generated by the switch.

According to Bifulco et al. (2016) and Michel et al. (2021), the InSP function, like any particular offloading function, faces the challenging issue of finding the relevant positioning with respect to the broad design space for delegation of control to SDN switches. In their opinion, a good approach to conceiving (eventually standardizing) a particular offloading function should involve a programming abstraction that achieves a fair compromise between viability and flexibility, far from extreme solutions that simply turn on well-known legacy protocol functions (e.g. MAC learning) or push a piece of code inside the switches (Tennenhouse and Wetherall 2007; Jeyakumar et al. 2013).

The authors of FOCUS (Yang et al. 2016) express the same challenges but, unlike the above proposals, they reject a performance-based design choice that requires adding new hardware primitives to OpenFlow switches in the development of the delegated control function. Instead, they promote a deployable software-based solution to be implemented in the switch's *software stack* to achieve a balanced trade-off between the flexibility and cost of the control function delegation process.

2.2.2. *Control plane distribution*

The second alternative, which we believe to be more effective, is to model the control plane in a way that mitigates scalability limitations (Abdullahi et al. 2021; Hohlfeld et al. 2018). In a physically centralized control model, a single SDN controller is in charge of handling all requests coming from SDN switches. As the network grows, the latter is likely to become a serious bottleneck in terms of scalability and performance (Qiu et al. 2017). On the other hand, a physically distributed control model uses multiple controllers that maintain a logically centralized network view. This solution is valued for handling the controller bottleneck, hence ensuring a better scale of the network control plane while decreasing control-plane latencies.

Even though the distributed control model is considered a scalable option when compared to the centralized control model, achieving network scalability while preserving good performance requires a relevant control distribution scheme that takes into account both the organization of the SDN control plane and the physical placement of the SDN controllers. In this context, we recommend a hierarchical organization of the control plane over a flat organization for increased scalability and improved performance. We also believe that the placement of controllers should be further investigated and treated as an optimization problem that depends on specific performance metrics (Heller et al. 2012).

Finally, by physically distributing the SDN control plane for scalability (and reliability, covered in section 2.3) purposes, it is worth mentioning that new kinds of challenges may arise. In particular, to maintain the logically centralized view, a strongly consistent model can be used to meet certain application requirements. However, as we discuss in section 2.4, a strongly consistent model may introduce new scalability issues. In fact, retaining strong consistency when propagating frequent state updates might block the state progress and cause the network to become unavailable, thus increasing switch-to-controller latencies.

2.3. Reliability

Concerns about reliability have been considered serious in SDN (Moazzeni et al. 2018). Indeed, the data-to-control plane decoupling has a significant impact on the reliability of the SDN control plane. In a centralized SDN-based network, the failure of the central controller may collapse the overall network. In contrast, the use of multiple controllers in a physically distributed (but logically centralized) controller architecture alleviates the issue of a single point of failure.

Despite not providing information on how a distributed SDN controller architecture should be implemented, the OpenFlow standard (since version 1.2) enables a switch to simultaneously connect to multiple controllers. This OpenFlow option allows each controller to operate in one of three roles (*master, slave, equal*) with respect to an active connection to the switch. Leveraging on these OpenFlow roles, which refer to the importance of controller replication in achieving a highly available SDN control plane, various resiliency strategies have been adopted by different fault-tolerant controller architectures. Among the main challenges faced by these architectures are control state redundancy and controller failover.

2.3.1. *Control state redundancy*

Controller redundancy can be achieved by adopting different approaches for processing network updates. In the active replication approach (Spalla et al. 2016), also known as state machine replication, multiple controllers process the commands issued by the connected clients in a coordinated and deterministic way in order to concurrently update the replicated network state. The main challenge of this method lies in enforcing a strict ordering of events to guarantee strong consistency among controller replicas. This approach for replication has the advantage of offering high resilience with an insignificant downtime, making it a suitable option for delay-intolerant scenarios. On the other hand, in passive replication, referred to as primary/backup replication, one controller (the *primary*) processes the requests, updates the replicated state and periodically informs the other controller replicas (the *backups*) about state changes. Despite offering simplicity and lower overhead, the passive replication scheme may create (controller and switch) state inconsistencies and generate additional delay when the primary controller fails.

Additional concerns that should be explored are related to the kind of information to be replicated across controllers. Existing controller platform solutions follow three approaches for achieving controller state redundancy (Fonseca and Mota 2017): state replication (Koponen et al. 2010; Botelho et al. 2014), event replication (Tootoonchian and Ganjali 2010; Katta et al. 2015) and traffic replication (Fonseca et al. 2013).

Moreover, control distribution is a central challenge when designing a fault-tolerant controller platform. The centralized control approach that follows the simple master–slave concept (Botelho et al. 2014; Katta et al. 2015) relies on a single controller (the *master*) which keeps the entire network state and takes all decisions based on a global network view. Backup controllers (the *slaves*) are used for fault tolerance purposes. The centralized alternative is usually considered in small- to medium-sized networks. On the other hand, in the distributed control approach (Berde et al. 2014; Ruslan et al. 2020; Koponen et al. 2010), the network state is partitioned across many controllers which simultaneously take control of the network while exchanging information to maintain the logically centralized network view. In this model, controller coordination strategies should be applied to reach agreements and solve the issues of concurrent updates and state consistency. Mostly effective in large-scale networks, the distributed alternative provides fault tolerance by redistributing the network load among the remaining active controllers.

Finally, the implementation aspect is another major challenge in designing a replication strategy (Spalla et al. 2016). While some approaches opted for replicating controllers that store their network state locally and communicate through a specific group coordination framework (Yazici et al. 2014), other approaches replicate the network state by delegating state storage, replication and management to external data stores (Berde et al. 2014; Koponen et al. 2010; Tootoonchian and Ganjali 2010) like distributed data structures and distributed file systems.

2.3.2. *Controller failover*

Apart from controller redundancy, other works have focused on failure detection and controller recovery mechanisms. Some of these works considered reliability criteria from the outset in the placement of distributed SDN controllers. Both the number and locations of controllers were determined in a reliability-aware manner while preserving good performance. Reliability was indeed introduced in the form of controller placement metrics (switch-to-controller delay, controller load) to prevent worst-case switch-to-controller re-assignment scenarios in the event of failures. Other works elaborated on efficient controller failover strategies which consider the same reliability criteria. Strategies for recovering from controller failures can be split into redundant controller strategies (with backups) and non-redundant controller strategies (without backups) (Kong 2017).

The redundant controller strategy assumes more than one controller per controller domain. One primary controller actively controls the network domain, and the remaining controllers (backups) automatically take over the domain in case it fails.

Despite providing a fast failover technique, this strategy depends on the associated standby methods (*cold*, *warm* or *hot*), which have different advantages and drawbacks (Pashkov et al. 2014). For instance, the cold standby method imposes a full initialization process on the standby controller given the complete loss of the state upon the primary controller failure. This makes it an adequate alternative for stateless applications. In contrast, the hot standby method is effective in ensuring a minimum recovery time with no controller state loss, but it imposes a high communication overhead due to the full state synchronization requirements between primary and standby controllers. The warm standby method reduces that communication overhead at the cost of a partial state loss.

On the other hand, the non-redundant controller strategy requires only one controller per controller domain. If it fails, controllers from other domains extend their domains to adopt orphan switches, thereby reducing the network overhead. Two well-known strategies for non-redundant controllers are the greedy failover and the pre-partitioning failover (Obadia et al. 2014). While the former strategy relies on neighbor controllers to adopt orphan switches at the edge of their domains and from which they can receive messages, the latter relies on controllers to proactively exchange information about the list of switches to take over in controller failure scenarios.

A number of challenges and key design choices based on a set of requirements are involved when adopting a specific controller replication and failover strategy. In addition to reliability and fault tolerance considerations, scalability, consistency and performance requirements should be properly taken into account when designing a fault-tolerant SDN controller architecture.

2.4. Controller state consistency

Contrary to physically centralized SDN designs, distributed SDN controller platforms face major consistency challenges (Schiff et al. 2016; Botelho et al. 2016; Zhang et al. 2018; Ahmad and Mir 2021). Clearly, physically distributed SDN controllers must exchange network information and handle the consistency of the network state being distributed across them and stored in their shared data structures in order to maintain a logically centralized network-wide view that eases the development of control applications. However, achieving a convenient level of consistency while maintaining good performance in software-defined networks facing network partitions is a complex task. As claimed by the CAP theorem applied to networks (Panda et al. 2013), it is generally impossible for SDN networks to simultaneously achieve all three of consistency (C), high availability (A) and partition tolerance (P). In the presence of network partitions, a weak level of consistency in exchange for high availability (AP) results in state staleness causing an incorrect behavior of applications, whereas a strong level of consistency serving the correct enforcement of network policies (CP) comes at the cost of network availability.

2.4.1. *Static consistency*

The *strong consistency* model used in distributed file systems implies that only one consistent state is observed by ensuring that any read operation on a data item returns the value of the latest write operation that occurred on that data item. However, such consistency guarantees are achieved at the penalty of increased data store access latencies. In SDNs, the strong consistency model guarantees that all controller replicas in the cluster have the most updated network information, albeit at the cost of increased synchronization and communication overhead. In fact, if certain data occurring in different controllers are not updated to all of them, then these data are not allowed to be read, thereby impacting network availability and scalability.

Strong consistency is crucial for implementing a wide range of SDN applications that require the latest network information and that are intolerant of network state inconsistencies. Among the distributed data store designs that provide strong consistency properties are the traditional SQL-based relational databases such as Oracle and MySQL.

On the other hand, as opposed to the strong consistency model, the *eventual consistency* model (sometimes referred to as a *weak consistency* model) implies that concurrent reads of a data item may return values that are different from the actual updated value for a transient time period. This model takes a more relaxed approach to consistency by assuming that the system will eventually (after some period) become consistent in order to gain network availability. Accordingly, in a distributed SDN scenario, reads of some data occurring in different SDN controller replicas may return different values for some time before eventually converging to the same global state. As a result, SDN controllers may temporarily have an inconsistent network view and thus cause incorrect application behavior.

Eventually consistent models have also been extensively used by SDN designers for developing inconsistency-tolerant applications that require high scalability and availability. These control models provide simplicity and efficiency of implementation but they push the complexity of resolving state inconsistencies and conflicts to the application logic and the consensus algorithms being put in place by the controller platform. Cassandra (Lakshman and Malik 2010), Riak (Klophaus 2010) and Dynamo (Sivasubramanian 2012) are popular examples of NoSQL databases that have adopted the eventual consistency model.

All things considered, maintaining state consistency across logically centralized SDN controllers is a significant SDN design challenge involving trade-offs between policy enforcement and network performance (Levin et al. 2012). The issue is that achieving strong consistency in an SDN environment that is prone to network failures is almost impossible without compromising availability and without adding complexity to network state management. Panda et al. (2013) proposed new ways to circumvent these impossibility results, but their approaches can be regarded as specific to particular cases.

2.4.2. *Adaptive multi-level consistency*

In a more general context, SDN designers need to leverage the flexibility offered by SDN to select the appropriate consistency models for developing applications with various degrees of state consistency requirements and with different policies. In particular, adopting a single consistency model for handling different types of shared states may not be the best approach to coping with such a heterogeneous SDN environment. In fact, recent works on SDN have stressed the need to achieve consistency at different levels. To date, two levels of consistency models have been applied to SDNs and adopted by most distributed SDN controller platforms: strong consistency and eventual consistency.

In our opinion, a hybrid approach that merges various consistency levels should be considered to find the optimal trade-off between consistency and performance. Unlike the approaches mentioned, which are based on static consistency requirements whereby SDN designers decide which consistency level should be applied for each knowledge upon application development, we argue that an SDN application should be able to assign a priority for each knowledge and, depending on the network context (i.e. instantaneous constraints, network load and so on), select the appropriate consistency level that should be enforced.

In this regard, recent approaches (Zhang et al. 2018; Aslan and Matrawy 2016; Sakic et al. 2017; Aslan and Matrawy 2018; Sakic and Kellerer 2018) introduced the concept of adaptive consistency in the context of distributed SDN controllers, whereby adaptively consistent controllers can tune their consistency level to reach the desired level of performance based on specific metrics. This alternative has the advantage of sparing application developers the tedious task of selecting the appropriate consistency level and implementing multiple application-specific consistency models. Furthermore, the approach can be efficient in handling the issues associated with eventual consistency models (Sakic et al. 2017).

Finally, in the same way as scalability and reliability, we believe that consistency should be considered when investigating the optimal placement of controllers. In fact, minimizing inter-controller latencies (distances), which are critical for system performance, facilitates controller communications and enhances network state consistency.

2.5. Interoperability

2.5.1. *Interoperability between the SDN controllers*

To foster the development and full adoption of SDN, we must overcome the common challenge of ensuring service interoperability between disparate distributed SDN controllers that belong to different SDN domains and use different controller technologies.

In today's multi-vendor environments, the limited interoperability between SDN controller platforms is mainly due to a lack of open standards for inter-controller communications. Aside from the standardization of the southbound interface – OpenFlow being the most popular southbound standard – to date there is no open standard for the northbound and east-westbound interfaces to provide compatibility between OpenFlow implementations.

Despite the emerging standardization efforts being carried out by SDN organizations, we argue that there are many barriers to effective and rapid standardization of the SDN east-westbound interfaces, including the heterogeneity of the data models being used by SDN controller vendors. Accordingly, we emphasize the need for common data models to achieve interoperability and facilitate the tasks of standardization in SDNs. In this context, YANG (Wallin and Wikström 2011) has emerged as a solid data modeling language used to model configuration and state data for standard representation. This NETCONF-based contribution from IETF is intended to be extended in the future and, more importantly, is expected to pave the way for the emergence of standard data models driving interoperability in SDN networks.

Among the recent initiatives taken in that direction, we can mention OpenConfig's effort on building a vendor-neutral data model written in YANG for configuration and management operations. Also worth mentioning is ONF's OF-Config protocol (OF-CONFIG 1.2: OpenFlow Management and Configuration Protocol 2014), which implements a YANG-based data model referred to as the Core Data Model. This protocol was introduced to enable remote configuration of OpenFlow-capable equipment.

2.5.2. *SDN interoperability with legacy networks*

Alongside the concerns about the interoperability between the diverse SDN controller implementations, we highlight another important SDN challenge that is often overlooked, namely, the challenge of reaching interoperability with legacy non-SDN technologies. While the deployment of SDN is fairly straightforward for new networks incorporating new SDN-ready devices, the transition from a legacy networking environment to SDN requires a period of co-existence between SDN and legacy technologies.

In such heterogeneous network architectures, operating a mix of SDN and traditional devices, it is extremely important to implement specific protocol mechanisms that support SDN control plane communications while providing efficient compatibility with existing IP control plane technologies. One potential solution is to adopt an incremental deployment strategy (Amin et al. 2018; Sandhya et al. 2017; Khorsandroo et al. 2021) according to which a few SDN-enabled devices are deployed in a traditional network among the legacy devices, incrementally replacing them, and forming the so-called hybrid SDN network. In such a network, both SDN and legacy nodes should operate in parallel and may communicate together in order to ensure an effective, gradual transition to SDN while reducing the associated operational costs and minimizing disruption of network services.

2.6. Other challenges

An efficient network monitoring is required for the development of control and management applications in distributed SDN-based networks. However, collecting the appropriate data and statistics without impacting the network performance is a challenging task. In fact, the continuous monitoring of network data and statistics may generate excessive overheads and thus affect network performance, whereas the lack of monitoring may cause an incorrect behavior of management applications. Current network monitoring proposals have developed different techniques to find the appropriate trade-offs between data accuracy and monitoring overhead. In particular, IETF's NETCONF southbound protocol provides some effective monitoring mechanisms for collecting statistics and configuring network devices. In the near future, we expect the OpenFlow specification to be extended to incorporate new monitoring tools and functions.

Similar to network monitoring, network security is another crucial challenge which should be studied. Decentralization of the SDN control reduces the risk associated with a single point of failure and attacks (e.g. the risk of a DDoS attack). However, the integrity of data flows between the SDN controllers and switches is still not safe. For instance, we can imagine that an attacker can corrupt the network by acting as an SDN controller. In this context, new solutions and strategies (e.g. based on TLS/SSL) have been introduced with the aim of guaranteeing security in SDN environments.

Another aspect related to SDN security is the isolation of flows and networks through network virtualization. In the case of an underlying physical SDN network, this could be implemented using an SDN network hypervisor that creates multiple logically isolated virtual network slices (called vSDNs), each of which is managed by its own vSDN controller (Blenk et al. 2016). At this point, care should be taken to design and secure the SDN hypervisor as an essential part of the SDN network.

2.7. Conclusion

While the need for a distributed SDN architecture has ultimately been recognized by the SDN community (Canini et al. 2014; Sarmiento et al. 2021; Canini et al. 2015), the best approach to designing and implementing an efficient (e.g. scalable and reliable) distributed SDN control plane is highly debatable given the many challenges presented by such distributed systems, as discussed above.

The scalability, reliability, consistency and interoperability of the SDN control plane are among the key challenges faced in designing an efficient and robust high-performance distributed SDN controller platform. Although regarded as the main limitations of fully centralized SDN control designs, scalability and reliability are also major concerns when designing a distributed SDN architecture. They are indeed heavily impacted by the structure of the distributed SDN control plane (e.g. flat, hierarchical or hybrid organization) as well as the number and placement of the multiple controllers within the SDN network. Achieving such performance and availability requirements usually comes at the cost of guaranteeing a consistent centralized network view that is required for the design and correct behavior of SDN applications. Consistency considerations should therefore be explored among the trade-offs involved in the design process of an SDN controller platform. Lastly, the interoperability between different SDN controller platforms of multiple vendors is another crucial operational challenge surrounding the development, maturity and commercial adoption of SDN. Overcoming that challenge calls for major standardization efforts at various levels of inter-controller communications (e.g. data models, northbound and east-westbound interfaces). Furthermore, such interoperability guarantees, with respect to different SDN technology solutions, represent an important step toward easing the widespread interoperability of these SDN platforms with legacy networks and, effectively ensuring the gradual transition toward softwarized network environments.

In the following chapters, we address the distributed SDN control problem by focusing on two major yet manageable challenges that, although correlated, could be treated as separate research problems: the controller placement problem (1) and the knowledge dissemination problem (2):

The first problem investigates the required number of SDN controllers along with their appropriate locations with respect to the desired objectives (see Chapter 3). The second problem addresses the type and amount of network information to be shared across the SDN controllers given a desired level of application state consistency and performance (see Chapters 4 and 5).

Scalability and Reliability Aware SDN Controller Placement Strategies

3.1. Introduction

In this chapter, we put forward novel strategies that address several aspects of the controller placement problem with respect to multiple reliability and performance criteria based on different uses and contexts (Bannour et al. 2017).

Our contribution to solving the controller placement problem is indeed intended for expanding IoT-like networks which face important scalability challenges in addition to reliability issues. The proposed SDN controller placement scheme uses heuristics with low computation time in order to deal with such large-scale and dynamic network environments where fast reevaluations of controller placement configurations are required to adapt in real-time to frequently changing network conditions. The potential of such heuristics in the context of SDN controller placement is explored by comparing two different types of heuristic-based algorithms according to various context-based strategies.

This chapter is organized as follows. In section 3.2 we provide an overview of state-of-the-art contributions that have addressed the controller placement problem. In section 3.3 we review the controller placement optimization problem and investigate the reliability and performance metrics involved. In section 3.4 we put forward our versatile approach to tackling this problem. In section 3.5 we present the obtained results. Finally, section 3.6 critically analyzes and discusses these results before elaborating on future perspectives.

3.2. Related work

Latterly, there has been growing interest in designing the distributed SDN control plane.

Heller et al. (2012) first motivated the SDN controller placement problem (CPP) and discussed the challenges of control plane reliability, scalability and performance.

The authors provided useful guidance on *how many* SDN controllers were needed and *where* they should be placed in order to achieve high performance in an SDN network. They argued that the optimal number of required SDN controllers must be planned carefully for each network topology, based on the concept of diminishing returns. They also claimed that in most topologies, one single controller is sufficient for fulfilling latency requirements but obviously insufficient for achieving control plane resilience.

In their study, the location and placement of a determined number of SDN controllers was treated as a variant of the *facility location problem*. Their placement strategy was only focused on minimizing the controller-to-switch propagation latency in the context of wide-area SDN deployments (e.g. Internet2), and analyzing the trade-offs between optimizing the average latency (the *k-median problem*) and the maximum latency (the *k-center problem*).

Their work has been extended by Hock et al. (2013) to incorporate other important performance aspects aside from the controller-to-switch latency in the multi-objective controller placement process, such as the resilience metrics with respect to controller failure, network disruption, load imbalance and inter-controller latency. In this context, Hock et al. introduced the resilient Pareto-based Optimal COntroller placement (POCO) optimization framework for providing all possible Pareto-optimal CPP solutions and finding the adequate trade-offs between quality, in terms of latency and resilience. Through the assessment of the framework using a range of different real network topologies, the authors argued that the required number of SDN controllers should be approximately 20% of all network nodes in order to meet resilience requirements.

While the first version of POCO was intended for small and medium-sized networks in which an exhaustive exploration of the entire solution space for selecting the optimal controller placement, with respect to the considered objectives, is computationally feasible, a subsequent version proposed by Lange et al. (2015b) comprised a *heuristic-based* Multi-Objective Combinatorial Optimization (MOCO) approach, namely, Pareto-simulated annealing (PSA), for dealing with the resilient controller placement problem in large-scale or dynamic network environments. However, when evaluating this approach on a set of real-world network topologies from the Internet Topology Zoo (Knight et al. 2011), in which the network size ranges between 5 and 50 nodes, the authors only emphasize the geographic extent aspect of large-scale networks and do not assess their scalability in terms of an increased number of network nodes.

Unlike the above strategies, which mitigate the impact of specific cases of network failures by minimizing resilience metrics such as the worst-case controller-to-switch latency, Hu et al. (2013) quantify reliability in terms of connectivity between the forwarding devices and their controllers (and between controllers as well) using a novel metric referred to as the *expected percentage of control path loss*. However, in doing so, the authors omitted important details about the failure probability of a network component, which, in our opinion, should be estimated based on a specific network failure model. Finally, different heuristic algorithms were presented and compared in their study for analyzing the trade-offs between latency and reliability in the reliability-aware controller placement decision.

Guang and Guo (2014) addressed another variant of the CPP that takes into account the load on controllers in addition to latency considerations in the placement strategy (the *capacitated k-center problem*). They used an effective algorithm for minimizing the maximum propagation delay under a controller capacity constraint. In their experiments, the load on controllers was measured based on the arriving rate of events, and the controller capacity was determined according to their access bandwidth. Their approach proved efficient in minimizing the required number of controllers to avoid controller overload.

Lange et al. (2015b) explored in Lange et al. (2015a) the potential of *specialized heuristics* to solve the capacitated variant of the multi-objective controller placement problem in large-scale SDN networks by developing the Pareto-capacitated k-medoids (PCKM) method, based on the k-medoids clustering algorithm. Such a specialized heuristic that optimizes case-specific criteria, namely, the average controller-to-switch latency and the controller load imbalance, was compared to generic heuristics usually destined for arbitrary multi-objective optimization purposes such as the MOCO PSA technique proposed in their previous work (Lange et al. 2015b). The performance comparison between these optimization heuristics applied to the CPP was also assessed on the Internet Topology Zoo (Knight et al. 2011) in terms of both the accuracy (with respect to the original Pareto frontier) and the runtime of the obtained solutions.

Similarly, Ahmadi et al. (2015) and Jalili et al. (2018) formulated the SDN CPP in highly dynamic or large-scale network environments as a MOCO problem and adapted an efficient multi-objective heuristic algorithm called the Non-dominated Sorting Genetic Algorithm (NSGA-II) to find a good and diverse approximation set of the Pareto-optimal front solutions, with respect to multiple competing criteria.

Their work provided an extensive analysis of the trade-offs between different combinations of the crucial SDN control plane resilience and performance metrics.

Another interesting work, which approaches the resilient SDN controller placement problem from a slightly different perspective, is presented in Müller et al. (2014). The authors propose a twofold controller placement scheme for improving the SDN control plane *survivability*. In contrast to previous works, which generally take the shortest path for granted when modeling connections between devices and controllers, the authors leverage the diversity of paths to place controllers at locations where the chance of controller-to-switch connectivity loss in the event of failures is minimized. More specifically, they formulate the placement of controller instances as an MILP problem to maximize the number of node disjoint paths between the controllers and the assigned switches under a controller capacity constraint. The second part of their contribution involved smart recovery mechanisms that use heuristics for selecting the optimal list of backup controllers for each forwarding device, based on both proximity and residual capacity considerations. A potential limitation of the resilience-oriented approach adopted by Müller et al. (2014) is that performance aspects are overlooked. Enhancing control plane connectivity may indeed come at the cost of generating high controller-to-switch delays.

Finally, in contrast to previous placement strategies that usually optimize the controller *locations* within the network given a fixed number of SDN controllers, works found in Ros and Ruiz (2016), Sanner et al. (2016), Perrot and Reynaud (2016) and Schütz and Martins (2020) place additional focus on minimizing the *number* of SDN controllers using different optimization strategies (e.g. heuristic-based approaches (Jalili et al. 2018; Schütz and Martins 2020; Singh et al. 2020), clustering techniques (Sanner et al. 2016; Wang et al. 2018; Yujie et al. 2020) and CPLEX solvers (IBM ILOG CPLEX Optimizer)) based on various placement constraints.

3.3. The SDN controller placement optimization problem

3.3.1. *Problem statement*

Ensuring a scalable and reliable distributed (but logically centralized) SDN control plane depends crucially on the placement of these physically distributed SDN controllers. More specifically, the so-called *controller placement problem* (Das et al. 2019; Shirmarz and Ghaffari 2021; Wu et al. 2020) consists of finding the required *number* and the appropriate *locations* of the SDN controllers (among the network nodes) that efficiently partition the network into several SDN controller domains to achieve the best trade-off between performance and reliability metrics (see Figure 3.1).

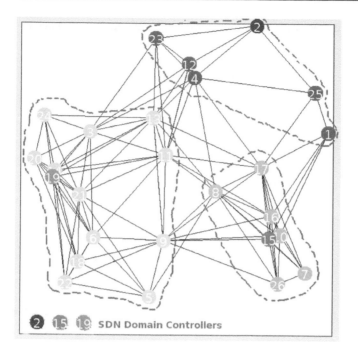

Figure 3.1. *The controller placement problem. For a color version of this figure, see www.iste.co.uk/bannour/software.zip*

3.3.2. *Problem formulation*

The network is viewed as a graph $G = (V, E)$, where the set of nodes V represents the network nodes comprising controllers and switches, while the set of edges E represents the links connecting these network nodes. Edge weights represent the shortest-path latencies between each pair of nodes. This information is stored in the available *Global Logical Network Topology Map* (see section 3.4.1), where $d(s, c)$ denotes the latency from a switch node $s \in V$ to a controller node $c \in V$. We formulate the controller placement problem as a multi-objective optimization problem according to a set of performance and reliability metrics (see Figure 3.2) discussed below.

3.3.3. *Placement metrics*

3.3.3.1. *Performance criteria*

Optimizing control plane performance is of paramount importance in large-scale IoT-like networks with stringent response-time requirements in which high propagation delays may lead to inconsistent and incorrect behaviors of network services.

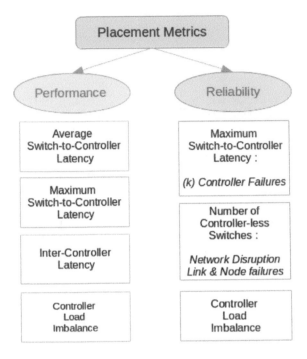

Figure 3.2. *Controller placement metrics. For a color version of this figure, see www.iste.co.uk/bannour/software.zip*

In particular, the average latency and the maximum latency between the switches and their associated controllers for a given placement C of k controllers among $n = |V|$ network nodes are two different latency-related performance metrics that were first introduced by Heller et al. (2012). Unlike the average latency placement metric (see equation [3.1]) which evaluates the overall quality of the network performance from a switch-to-controller latency point of view while hiding single cases of unacceptably high latencies, the maximum latency placement metric (see equation [3.2]) is indeed useful in preventing the occurrence of such high-latency cases in placement scenarios.

Average switch-to-controller latency:

$$\pi^{Avg-s2c-Latency}(C) = \frac{1}{n} \sum_{(s \in S)} \min_{(c \in C)} d(s,c) \qquad [3.1]$$

Maximum switch-to-controller latency:

$$\pi^{Max-s2c-Latency}(C) = \max_{(s \in S)} \min_{(c \in C)} d(s, c) \qquad [3.2]$$

Other important considerations that have a direct impact on the SDN control-plane performance include the inter-controller latencies (Lange et al. 2015b; Killi and Rao 2019). Physically distributed SDN controllers should indeed be placed as close as possible to each other in order to reduce the cost of maintaining a consistent logically centralized network view, i.e. the inter-controller communication and the global state synchronization. Accordingly, SDN controller locations can be selected in a way that minimizes the average and the maximum inter-controller latencies defined in [3.3] and [3.4] alongside the previously mentioned control-to-data plane performance metrics.

Inter-controller latencies:

$$\pi^{Avg-c2c-Latency}(C) = \frac{1}{|C|} \sum_{(c_1, c_2 \in C)} d(c_1, c_2) \qquad [3.3]$$

$$\pi^{Max-c2c-Latency}(C) = \max_{(c_1, c_2 \in C)} d(c_1, c_2) \qquad [3.4]$$

Aside from the above placement metrics, which affect the network performance from a switch-to-controller or controller-to-controller latency perspective, the controller capacity-awareness is another important performance factor that should be considered in the controller placement process in order to avoid the chance of controller overload and thereby prevent the related performance issues (additional delays at the controller level, etc.).

One way of tackling the controller overload aspect is by assuming that all controllers C have equal capacities Q_c (in terms of the number of controlled nodes in our case) and by guaranteeing an equal distribution of the control plane load (the sum of each load $l(s)$ to control switch s (Guang and Guo 2014)) among these controller instances [3.5]. Each controller should be loaded at 80% of its full capacity Q_c, leaving a controller capacity margin of 20% to prevent occasional controller overload. This fair load distribution is achieved by implementing an intelligent, well-balanced switch-to-controller assignment method that, given a fixed number of controller instances, assigns each network node to the closest controller provided the load on that controller does not exceed the imposed load constraint.

Moreover, the implemented controller assignment heuristic guarantees that any network node that could not be assigned to its closest controller due to the controller capacity constraint is intended for assignment to the second closest controller, which has not yet reached its full load capacity.

The controller load constraint:

$$\sum_{(s \in S)} l(s) = 80\% Q_c, \forall c \in C \tag{3.5}$$

An alternative controller load balancing scheme is to keep the usual shortest-path-based switch-to-controller assignment method, relax the fair load balancing constraint and instead introduce an additional load imbalance metric to be minimized through the controller placement optimization [3.6]. This metric is defined by Hock et al. (2013) as the difference between the maximum and the minimum number of network nodes assigned to a controller for a given controller placement C.

Load imbalance:

$$\pi^{Load-Imbalance}(C) = \max_{(c \in C)} n_c - \min_{(c \in C)} n_c \tag{3.6}$$

3.3.3.2. *Reliability criteria*

Although it provides several benefits in terms of increased flexibility and better performance, the physical SDN control-to-data plane separation feature introduces additional concerns in terms of network reliability as a crucial requirement for operational SDNs. In fact, one key consideration in the design of distributed SDN networks is to improve the reliability of the SDN control plane. This aspect of SDN reliability can be ensured by placing SDN controllers in a reliability-aware manner that mitigates the impact of controller failures. The most common reliability mechanism used for guarding against the failure of primary controllers is the assignment of the associated network switches to the closest working controllers. By doing so, response-time requirements should be satisfied in order to guarantee controller fault tolerance. In other words, the propagation latencies of these previously controlled switches, with respect to the new backup controllers, should remain acceptable.

As an indicator of reliability against controller instance failures, we use the maximum latency metric (to be minimized), which is computed based on the propagation latencies between the network switches and all the subsets of working controllers C_1 for a placement C according to the considered controller failure scenarios F, as defined in the general formula below:

$$\pi_F^{Max-s2c-Latency}(C) = \max_{(s \in S)} \max_{(C_1 \subseteq C)} \min_{(c \in C_1)} d(s, c) \tag{3.7}$$

Among these controller failure scenarios, the worst-case scenario for a network switch would be the simultaneous failure of the $(k - 1)$ closest SDN controllers.

Mitigating this control plane failure scenario implies minimizing the maximum of the latencies between network switches s and their respective furthest functional controllers $C_{Fu}(s)$ as follows:

$$\pi_{F(k-1)}^{Max-s2c-Latency}(C) = \max_{(s \in S, c \in C_{Fu}(s))} d(s,c) \qquad [3.8]$$

In practice, it is more common for primary controller failures to occur one at a time. Therefore, reducing the impact of this controller failure scenario implies minimizing the maximum of the latencies between network switches s and their respective second closest controllers $C_{Cl}(s)$, as expressed in the following:

$$\pi_{F(1)}^{Max-s2c-Latency}(C) = \max_{(s \in S, c \in C_{Cl}(s))} d(s,c) \qquad [3.9]$$

3.4. The proposed SDN controller placement scheme

3.4.1. *The adopted approach*

In this section, we present our two-phase approach to modeling and tackling the controller placement problem using a decentralized simulation framework. First, we deploy monitoring (data-gathering) mechanisms in order to gather and transmit information about the network topology. This collected information is then used by the controller placement optimization algorithms that we implemented in the second phase of the work.

For a given network, we start by running a distributed leader election scheme. The network nodes communicate with their neighboring nodes by sending leader request messages and then waiting for leader responses. In the meantime, nodes that did not receive a leader reply message may declare themselves leaders depending on a given leader election probability. This task ensures that each network node is managed by one leader node and that each elected leader node will assume responsibility for some part of the network.

Once the leader election process is complete, the network nodes start sending messages in order to record the desired information about their connected neighbors in a *Neighbor Map* (latency information in our case). At this point, all follower nodes send their neighbor maps to their respective leaders. In this way, leader nodes get the cluster information required to construct their *Local Leader Maps*.

Finally, leaders synchronize their local cluster information and build the *Global Physical Network Topology Map*. Among the set of network leader nodes, only one is nominated as the *Hyper Leader Node*, which will be responsible for running the Dijkstra's shortest path algorithm and building the *Global Logical Network Topology Map*.

At the hyper node level, controller placement optimization algorithms are implemented and run based on this available global network view and based on a determined number of network controllers k. Controller placement solutions are then investigated and analyzed in order to find the optimal trade-off between the considered reliability and performance metrics.

3.4.2. *Multi-criteria placement algorithms*

In order to optimize the placement of k SDN controllers according to the discussed performance and reliability metrics, we use two different algorithms, a clustering algorithm based on PAM (Partitioning Around Medoids) and a modified genetic algorithm called NSGA-II (Non-dominated Sorting Genetic Algorithm II).

PAM (Ganatra 2012) is a k-medoid clustering technique that partitions the data set of N objects (N network nodes) into k clusters represented by k medoids (the SDN controller nodes). The main idea of PAM is to find the optimal set of medoids that improves the overall quality of clustering, which is measured based on the average dissimilarity of all data objects to their nearest medoid. In our case, all the considered metrics M are of equal importance, thereby making the dissimilarity function D (to be minimized) for a given placement $C \in CP$ (the considered placement configurations) computed as the normalized sum of all weighted objectives O with the associated weights equal to $\frac{1}{M}$ as follows:

$$D^{PAM-B}(C) = \sum_{i \in M} (\frac{1}{M}) \times N(O_i) \qquad [3.10]$$

where:

$$N(O_i) = \frac{O_i(C) - \min_{(C \in CP)} O_i(C)}{\max_{(C \in CP)} O_i(C) - \min_{(C \in CP)} O_i(C)}$$

Algorithm 1, called PAM-B, corresponds to the multi-criteria controller placement algorithm that we developed based on PAM.

Algorithm 1 PAM-B:

1: n nodes, an integer k.
2: Init: Select k nodes at random and define them as medoids.
3: Associate each object to the appropriate medoid according to a well-defined assignment method.
4: **for** each medoid m **do**
5:    Compute and store the objective function values of the current configuration.
6:    **for** each non-medoid r **do**
7:       Swap m and r
8:       Associate each object to the appropriate medoid according to the considered assignment method.
9:       Compute and store the objective function values of the new configuration.
10:    **end for**
11:    Compute the maximum and the minimum values of each objective function over the considered configurations.
12:    Compute the normalized total dissimilarity of each considered configuration based on the above optima.
13:    Select the configuration with the lowest normalized total dissimilarity
14: **end for**

On the other hand, NSGA-II (Ahmadi et al. 2015) is a popular, fast and elitist genetic algorithm for multi-objective optimization. In addition to the classical genetic operators (crossover and mutation), NSGA-II uses other multi-objective ranking mechanisms (non-dominated sorting and the crowding distance) for creating the next-generation population of candidate solutions. The main idea behind NSGA-II is to make this population evolve toward a set of optimal non-dominated solutions (the Pareto front) representing the best trade-offs between the considered objectives. In this work, we set a list of NSGA-II parameters as follows:

Parameters	Values
Population	$k * 2$
Selection operator	sbx
MaxEvaluations	Depends on the strategy (see Table 3.2)

Table 3.1. *NSGA-II parameters*

3.4.3. *Gradual strategies*

We propose multiple strategies for tackling the SDN controller placement problem according to our performance and reliability criteria. We follow a step-by-step approach based on the gradual incorporation of these placement metrics for assessment by our multi-criteria algorithms (PAM-B and NSGA-II). Thus, it becomes possible to investigate the direct impact of these placement metrics on the quality of the controller placement solutions and also to make the controller placement approach adaptable to various use cases. More importantly, such a versatile approach can be leveraged by SDN operators to assist them in finding their optimal controller placement solution, tailored to their specific context.

Strategy 1: a latency-based strategy

This strategy solves the SDN controller placement problem based on the two latency-related performance metrics shown in [3.1] and [3.2] while keeping the usual shortest-path switch-to-controller assignment method. Accordingly, the multi-objective NSGA-II is launched with these two objectives to be minimized, while PAM-B minimizes the following dissimilarity function as a normalized sum of the two considered objectives, in accordance with [3.10]:

$$D_1^{PAM-B}(C) = (\frac{1}{2}) \times N(\pi^{Avg-Latency}(C)) + (\frac{1}{2}) \times N(\pi^{Max-Latency}(C))$$

[3.11]

Strategy 2: Strategy 1 under a load capacity constraint

In addition to the previously mentioned latency-related performance metrics, Strategy 2 incorporates a fair load balancing scheme by turning the simple switch-to-controller assignment method of Strategy 1 into an intelligent assignment method, which guarantees an equal distribution of the control plane load among controller instances (80% of their equal capacities) and, at the same time, a fair affectation of network switches to their closest lightly loaded controllers (see section 3.3.3.1). Thus, in Strategy 2, PAM-B minimizes the same dissimilarity function [3.11] used in Strategy 1.

Strategy 3: Strategy 1 with a load imbalance metric

In Strategy 3, along with the performance metrics of Strategy 1, we adopt an alternative load balancing scheme using the load imbalance metric [3.6] proposed by Hock et al. (2013) and we investigate the controller overload risk.

The following formula defines the dissimilarity function of PAM-B based on the three considered objectives:

$$D_3^{PAM-B}(C) = (\frac{1}{3}) \times N(\pi^{Avg-Latency}(C)) + (\frac{1}{3}) \times N(\pi^{Max-Latency}(C))$$

$$+ (\frac{1}{3}) \times N(\pi^{Load-Imbalance}(C)) \quad [3.12]$$

Strategy 4: Strategy 3 with reliability metrics

Strategy 4 provides a rich SDN controller placement optimization framework which includes reliability metrics (explained in section 3.3.3.2) along with performance metrics. For reliability placement metrics [3.7], users of the framework have the option of including a reliability metric variant, which tackles the worst-case controller failure scenario [3.8], or a variant that addresses a more common controller failure scenario [3.9], in addition to the previous performance metrics [3.1], [3.2] and [3.6].

The dissimilarity functions of PAM-B for both variants are calculated in accordance with the following formulas:

$$D_4^{PAM-B(k-1)}(C) = (\frac{1}{4}) \times N(\pi^{Avg-Latency}(C)) + (\frac{1}{4}) \times N(\pi^{Max-Latency}(C))$$

$$+ (\frac{1}{4}) \times N(\pi^{Load-Imbalance}(C)) + (\frac{1}{4}) \times N(\pi_{F(k-1)}^{Max-Latency}(C)) \quad [3.13]$$

$$D_4^{PAM-B(1)}(C) = (\frac{1}{4}) \times N(\pi^{Avg-Latency}(C)) + (\frac{1}{4}) \times N(\pi^{Max-Latency}(C))$$

$$+ (\frac{1}{4}) \times N(\pi^{Load-Imbalance}(C)) + (\frac{1}{4}) \times N(\pi_{F(1)}^{Max-Latency}(C)) \quad [3.14]$$

3.5. Performance evaluation

3.5.1. *Simulation settings*

We use the JAVA-based distributed simulation framework Sinalgo (Simulator for Network Algorithms) for implementing our two-phase approach (explained in section 3.4.1) and evaluating our multi-criteria SDN controller placement algorithms (see section 3.4.2) according to gradual strategies and various scenarios (see section 3.4.3).

Table 3.2 summarizes the values corresponding to the *maximum number of objective function evaluations* simulation parameter, used as a stopping criterion in the NSGA-II algorithm as a function of the number of objectives involved in each strategy and the size of the network in each simulation scenario.

Number of objectives	Number of nodes	MaxEvaluations
2	20, 60, 100, 200, 400	10,000
(Strategies 1 and 2)	500, 600	20,000
	700	40,000
	800	60,000
	900	80,000
	1000	100,000
3	20, 60, 100, 200, 400	20,000
(Strategy 3)	500, 600	40,000
	700	60,000
	800	80,000
	900	100,000
	1000	120,000
4	20, 60, 100, 200, 400	40,000
(Strategy 4)	500, 600	60,000
	700	80,000
	800	100,000
	900	120,000
	1000	140,000

Table 3.2. *The maximum number of objective function evaluations (MaxEvaluations)*

3.5.2. *Simulation results*

In this section, we present the simulation results of the proposed approach based on the gradual strategies being followed and according to different simulation scenarios. For each strategy, and for a given network topology, we evaluate the controller placement solutions proposed by PAM-B and NSGA-II. PAM-B generates the optimal controller placement clustering solution with respect to the equally weighted dissimilarity measure defined in [3.10], in which we give equal importance to the considered objectives.

Similarly, for the multi-objective NSGA-II, we consider the fairest controller placement solution (in relation to the desired criteria) among all the generated, non-dominated Pareto-optimal solutions representing the possible trade-offs between the considered objectives. This is achieved by selecting the Pareto placement solution S that best reduces the total gap between all the associated objectives M and their respective optimal values across the set of all Pareto optimal solutions P. In our case, this implies considering the Pareto solution with the minimum value of the following measure [3.15]:

$$a(S) = \sum_{i \in M} (\frac{1}{M}) \times \frac{O_i(S) - \min_{(S \in P)} O_i(S)}{\max_{(S \in P)} O_i(S) - \min_{(S \in P)} O_i(S)} \qquad [3.15]$$

Accordingly, several simulation scenarios are performed following the considered strategies and using various types of network topologies of different size, from 20 to 1000 network nodes. This allowed us to compare our controller placement strategies, analyze the performance of both algorithms for solving the controller placement optimization problem and study the scalability of our approach, which is mainly intended for large-scale IoT-like deployments.

When comparing the optimal controller placement solutions across Strategies 1–3 with respect to the considered latency-based performance metrics (see Figure 3.3), we notice that, unlike Strategies 1 and 3 which show similar performance trends, Strategy 2 yields poorer results in minimizing both the average latency (Figure 3.3(a)) and the maximum latency (Figure 3.3(b)) performance metrics due to the imposed load balancing constraint. For example, in Scenario no. 12 (Figure 3.3(b)), in which the network size is equal to 1,000 nodes, both PAM-B and NSGA-II provided, according to Strategy 2, controller placement configurations in which the maximum latency value is above 400 ms compared to less than 100 ms for both Strategies 1 and 3. On the other hand, we note that in all 12 of the simulation scenarios considered in these three strategies, PAM-B is evidently better than NSGA-II at simultaneously minimizing both the average latency and the maximum latency performance metrics of the obtained controller placement configurations. For instance, when it comes to the average latency metric, PAM-B according to Strategy 1 is better (from 6% to 40%) than NSGA-II over all scenarios, (from 10% to 50%) according to Strategy 2 and (up to 20%) according to Strategy 3.

However, the obvious advantage of Strategy 3, which adds a load imbalance metric to Strategy 1 (Figure 3.4(a)), over Strategy 2, which incorporates a load balancing constraint in Strategy 1, is the fact that it did not deteriorate the level of latency-based performance targeted by Strategy 1. However, the potential drawback of Strategy 3 is related to the risk of controller overload, as illustrated by Figure 3.4(b), which depicts the percentage of overloaded controllers in the considered scenarios. For instance, in Scenario no. 10 (Figure 3.4(b)), in which the network size is equal to 800 nodes and the number of controllers is equal to 80, both PAM-B and NSGA-II produced controller placement configurations in which 21 (26.25%) of these controllers are overloaded.

As explained in section 3.4.3, Strategy 4 involves reliability metrics in addition to the set of performance metrics considered by Strategy 3. In particular, Figure 3.5, for each variant of Strategy 4 according to all the optimal controller placement solutions, compares the values of the maximum latency metric in the failure-free case with that of the maximum latency metric in the considered failure case scenario. It also shows that PAM-B and NSGA-II perform in a similar way when optimizing these metrics.

Optimal Controller Placement

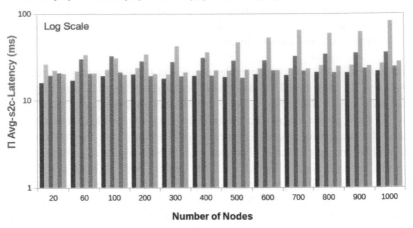

a)

Optimal Controller Placement

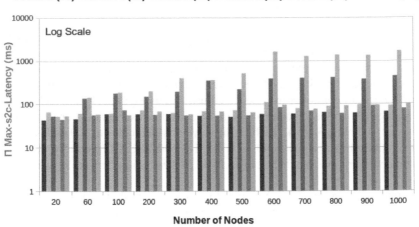

b)

Figure 3.3. *Strategies 1–3: latency-based performance metrics. For a color version of this figure, see www.iste.co.uk/bannour/software.zip*

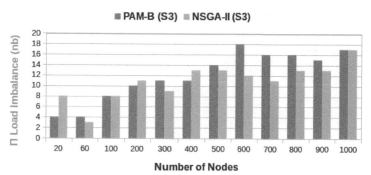

a)

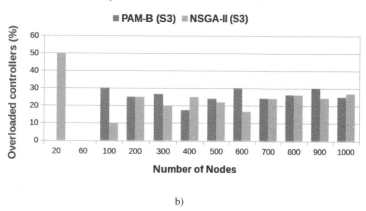

b)

Figure 3.4. *Strategy 3: load imbalance. For a color version of this figure, see www.iste.co.uk/bannour/software.zip*

Figure 3.6 analyzes the performance cost of taking into account reliability criteria in the controller placement optimization process. Surprisingly, optimizing for reliability metrics did not severely impact performance metrics like the maximum latency (in the failure-free case) (Figure 3.6(a)) whose values remained acceptable and comparable to those in Strategy 3 except for a few placement scenarios that were in most cases produced by NSGA-II. In addition, similar trends can be observed across Strategies 3 and 4 for each of the load imbalance (Figure 3.6(b)) and the average latency (Figure 3.6(c)) performance metrics of the obtained placement configurations. For example, in Scenario no. 12 (Figure 3.6(c)), PAM-B(k-1)

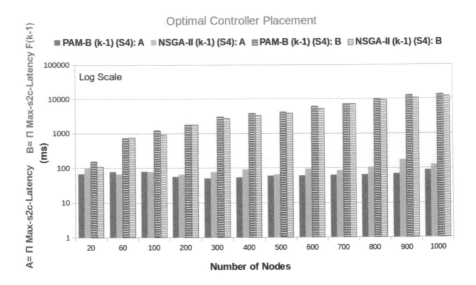

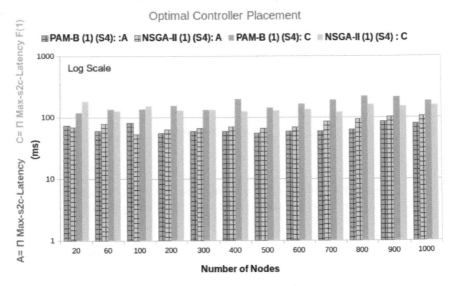

Figure 3.5. *Strategy 4: reliability metrics (maximum latencies in failure-free and failure case scenarios). For a color version of this figure, see www.iste.co.uk/bannour/software.zip*

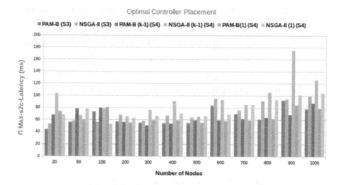

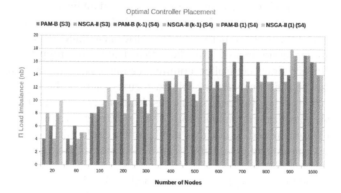

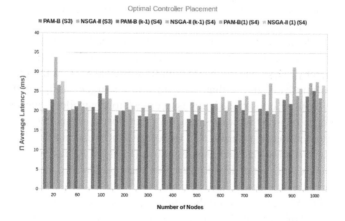

Figure 3.6. *Strategy 4: performance metrics. For a color version of this figure, see www.iste.co.uk/bannour/software.zip*

(respectively, PAM-B(1)) according to Strategy 4 produced an optimal controller placement configuration in which the value of the average latency metric is equal to 25.3 ms (respectively, 23.5 ms) compared to 24 ms for PAM-B according to Strategy 3. In the same scenario, NSGA-II(k-1) (respectively, NSGA-II(1)) according to Strategy 4 generated a controller placement configuration with an average latency value equal to 27.6 ms (respectively, 26.8 ms) against 27.3 ms for NSGA-II according to Strategy 3.

3.6. Discussion

Four strategies were put forward to address certain important aspects of the SDN controller placement problem. The first strategy centered on the optimal placement of SDN controllers based on both the average and the maximum latency criteria. Optimizing such latency-related performance metrics guarantees proper shortest-path switch-to-controller assignments. On the other hand, the proper placement of controllers, whereby inter-controller communication costs are optimized, means taking into account additional latency-related metrics such as the latencies between the individual controllers. However, such considerations are beyond the scope of this chapter; inter-controller communication effects (Zhang et al. 2016; Das and Gurusamy 2020) are indeed issues that need further exploration in our subsequent work about the knowledge dissemination part of distributed SDN control.

The second and third strategies were motivated by the observation that optimizing the locations of controllers based solely on latency-related metrics, as in Strategy 1, may generate placement configurations in which some controllers are in charge of a large number of network switches, and thus highly exposed to potential overload risks. Imposing a load balancing constraint complemented by an intelligent switch-to-controller assignment method (Strategy 2) guarantees a fair distribution of the control plane load, whereby SDN controllers are equally loaded at 80% of their total capacity. However, it is intuitively likely that, despite this intelligent assignment technique, some switches will be constrained for assignment to controllers that are relatively far (from a latency point of view) because all their closest controllers have somehow reached the imposed load constraint. On the other hand, relaxing this load constraint and substituting it with a load imbalance metric (Strategy 3) has proved effective in providing switches with better freedom to join their preferred controller cluster, but less immune to the risk of controller overload (see Figure 3.4(b)). A potential solution to addressing this problem could be the implementation of a heuristic method, to be launched at the end of Strategy 3, in order to cope with such controller overload cases. One way of doing this is to turn, in each overloaded controller cluster, a certain number of switches (the closest to the cluster controller) into additional controllers that will handle the extra controller cluster overload. For each overloaded cluster, the number of switches to become controllers may depend on the surplus number of cluster switches above the cluster controller capacity. In all the scenarios considered by Strategy 3, the maximum controller load has never exceeded 200% of its total capacity, thereby confirming the need for no more than one additional controller in each overloaded cluster.

Considering controllers with different capacities, as in real network settings, is another important factor to take into account. Another controller placement strategy can be proposed and implemented using a load balancing scheme based on different node capacities.

The fourth strategy incorporated reliability metrics in addition to the considered performance metrics. For example, the reliability-aware controller placement, which takes into account worst-case failure scenarios, seems to impact the locations of controllers in a way that places them closer to the network center to minimize worst-case latencies with respect to all network switches and thus to ensure an optimized switch-to-controller backup reassignment that preserves performance in cases of primary controller outages. However, aside from controller failures, other failure scenarios such as the failure of network switches and links may occur and can therefore be involved in the controller placement decision-making process in order to enhance the reliability of the SDN control plane.

Evaluating PAM-B and NSGA-II over the proposed strategies revealed that, in most scenarios, PAM-B outperforms NSGA-II in terms of the quality of final solutions with respect to the considered metrics (see section 4.6.2). More specifically, PAM-B gives a more balanced trade-off between performance and reliability metrics and, more importantly, it produces more stable results over all strategies, whereas the performance of NSGA-II with respect to these metrics is sometimes unpredictable and highly dependent on the strategy being followed.

Finally, implementing different kinds of heuristic-based algorithms (PAM-B and NSGA-II) directed at solving the controller placement problem within a reasonable time frame, and testing them over large network instances, demonstrated the scalability of the proposed approach and its adequacy with the IoT context. In particular, Figure 3.7 illustrates the runtime comparison between PAM-B and NSGA-II, which reflects similar trends over the considered scenarios. In fact, the computational complexities of PAM-B and NSGA-II are close and equal to $O(k(n-k)^2)$ (k is the number of medoid clusters – the number of SDN controllers in our case – and n is the number of objects; the network size in our case) and $O(MN^2)$ (M is the number of objective functions and N is the population size).

However, it is worth mentioning that the computation time of clustering approaches such as PAM can be significantly improved. For instance, CLARA (Clustering LARge Applications) (Shraddha and Emmanuel 2014), a sampling-based variant of PAM, is highly recommended for dealing efficiently with large data sets. It indeed has a computational complexity of $O(ks^2 + k(n-k))$, where k is, in our case, the number of SDN controllers, n is, in our case, the network size and s is the sample size. In fact, a CLARA-based approach can be used instead of PAM-B in large-scale network scenarios in order to further reduce the overall computation time.

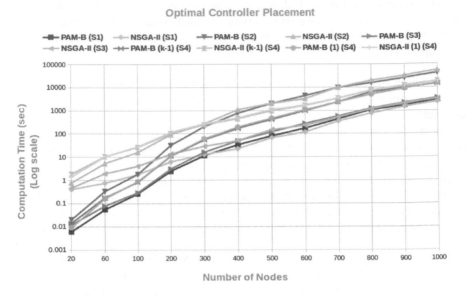

Figure 3.7. *Computation time comparison between PAM-B and NSGA-II over the considered strategies. For a color version of this figure, see www.iste.co.uk/bannour/software.zip*

3.7. Conclusion

We investigated the SDN controller placement issue in the context of large-scale IoT-like networks and underlined the need for an efficient approach to this multi-objective optimization problem. In this respect, several SDN control plane performance and reliability metrics were considered according to different needs and contexts. Through these strategies, two heuristic approaches were proposed to find high-quality approximate solutions to the controller placement problem in a reasonable computation time: a clustering approach (PAM-B) based on a dissimilarity score and a modified genetic approach (NSGA-II). Our results demonstrate the potential of clustering techniques in delivering appropriate controller placement configurations that achieve balanced trade-offs between the competing performance and reliability criteria.

The challenge of determining the required number and locations of SDN controllers represents one particular aspect of the overall process of addressing the distributed SDN control problem. This leads us to the second key aspect which calls for further investigation, namely the knowledge sharing challenge associated with such logically centralized distributed SDN platforms. However, given the potential correlation between placing multiple SDN controllers and modeling the type of communication among them, it becomes essential to reevaluate certain factors involved in solving the controller placement problem after studying the variety of data consistency models for inter-controller communication in the context of physically distributed SDN architectures. It is also planned to implement and validate the approaches evaluated in simulation on an experimental SDN test bed based on OpenvSwitch nodes.

Adaptive and Continuous Consistency for Distributed SDN Controllers: Anti-Entropy Reconciliation Mechanism

4.1. Introduction

Logically centralized but physically distributed SDN controllers are mainly used in large-scale SDN networks for scalability, performance and reliability reasons. These controllers host various applications that have different requirements in terms of performance, availability and consistency. Current SDN controller platform designs employ conventional strong consistency models so that the SDN applications running on top of the distributed controllers can benefit from strong consistency guarantees for network state updates.

However, in large-scale deployments, ensuring strong consistency is usually achieved at the cost of generating performance overheads and limiting system availability. This makes weaker optimistic consistency models, such as the eventual consistency model, more attractive for SDN controller platform applications with high availability and scalability requirements. In this chapter, we argue that use of the standard eventual consistency models, though a necessity for efficient scalability in modern SDN systems, provides no bounds on the state inconsistencies tolerated by the SDN applications.

To remedy this, we propose an adaptive multi-level consistency model for the distributed ONOS controllers following the notion of continuous and compulsory (Yu and Vahdat 2000) eventual consistency, in which network application states adapt their eventual consistency level dynamically at runtime, based on the observed state inconsistencies under changing network conditions. This model presents many advantages when compared to the strong consistency and eventual consistency extremes, especially in large-scale deployments.

Our scalable consistency adaptation strategy was implemented for a source-routing application on top of the distributed open-source ONOS controllers. It mainly consists of turning ONOS's eventual consistency model into an adaptive consistency model using the *anti-entropy reconciliation period* as a *control knob* for an adaptive fine-grained tuning of consistency levels. Compared to ONOS's static state consistency management scheme at scale, our consistency strategy is aimed at minimizing state synchronization overheads while taking into account the application's continuous consistency service-level agreements (SLAs) (e.g. *numerical error* bounds) and without compromising the application requirements of high availability.

This chapter is organized as follows. In section 4.2 we provide an overview of the consistency models used by state-of-the-art SDN controller platforms. In section 4.3 we review the consistency problem in SDN and investigate the involved consistency trade-offs. In section 4.4 we discuss the consistency models implemented in the ONOS controller platform. In section 4.5 we present our consistency adaptation approach for the distributed ONOS controllers. Finally, section 4.6 presents and discusses the experimental results.

4.2. Related work

The challenges related to consistency in distributed SDN control have been recently addressed in the SDN literature. Some works focused on the impact of switch-to-controller state consistency *between switches and controllers* on network application performance. Reitblatt et al. (2011) studied the consistency of controller-driven flow updates in terms of network policy conservation. They proposed a new type of consistency abstraction to enforce a consistent forwarding state at different levels (per-flow consistency and per-packet consistency). More recent approaches (Nguyen et al. 2017) focused on efficiently updating the network data plane state while preventing forwarding anomalies at the switches and maintaining desired consistency properties (e.g. loop and black-hole freedom).

Another set of works, falling within the scope of this chapter, focused on achieving controller-to-controller state consistency *between the distributed controllers* without compromising application performance.

Current implementations of distributed SDN controller platforms offer different state consistency abstractions. They use static mono-level (ODL) or multi-level (ONOS) consistency models such as the strong, eventual and weak state consistency levels.

Onix (Koponen et al. 2010) offers two separate dissemination mechanisms for synchronizing network state updates between the NIBs stored at the controller instances. These mechanisms are based on two implemented data store options: a replicated transactional database designed for ensuring strong consistency at the cost of good performance for persistent but slowly changing states, and a high-performance memory-only distributed hash table (DHT) for volatile states that are tolerant to inconsistency.

Similarly, ONOS provides different state-sharing mechanisms to achieve a consistent network state across the cluster of ONOS controllers. More specifically, ONOS's distributed core eases the state management and coordination tasks for application developers by providing them with an available set of core building blocks for dealing with different types of distributed control plane states, including a consistent primitive for state requiring strong consistency and an eventually consistent primitive for state tolerating relaxed consistency.

On the other hand, ODL supports a strong consistency model in its distributed data store architecture. In fact, all the data shared across the cluster of controllers for maintaining the logically centralized network view is handled in a strongly consistent manner using the RAFT consensus algorithm (Ongaro and Ousterhout 2014).

More recent approaches to handling the issues of controller state consistency (Aslan and Matrawy 2016; Sakic et al. 2017, 2018) recommend the use of adaptive consistency for the distributed SDN controller platforms. Aslan and Matrawy (2016) attempted to mitigate the impact of controller state distribution on SDN application performance by proposing an adaptive tunable consistency model following the delta consistency model. In their model, the automatic control plane adaptation module tunes the consistency level (the synchronization period parameter) based on an application-specific performance indicator that is measured given the current state of the network. To assess their approach, the authors compared the performance of the distributed load-balancing application running on top of adaptive and non-adaptive controllers.

Aslan and Matrawy (2018) studied the feasibility of employing adaptive controllers that are built on top of tunable consistency models similar to that of Apache Cassandra. They presented an adaptation strategy for the SDN controllers that uses clustering techniques to map a given application performance indicator into an appropriate consistency level that can be used to configure the parameters associated with the underlying tunable consistency model. However, the authors did not test the validity of their proposal using a specific SDN application running on top of an SDN controller platform.

Similarly, Sakic et al. (2017) put forward an adaptive consistency model for distributed SDN controllers following the eventual consistency level. The main aim of changing the controller consistency level on-the-fly was to maintain a scalable system that sacrifices application optimality for less synchronization overhead. Accordingly, the authors propose a cost-based approach that bounds the correctness to a tunable threshold, where the consistency level is adapted based on the effort of state convergence after the expiration of a non-synchronization period and the application inefficiencies due to operations with stale state. The performance of the proposed model was evaluated based on a specific routing application.

4.3. The consistency problem in SDN

4.3.1. *Consistency trade-offs in SDN*

In distributed SDN architectures, the SDN control plane supports the interaction between multiple controllers through their "east-west" interfaces. Inter-controller communications are indeed needed to synchronize the controllers' shared data structures to maintain a consistent global network view, and therefore ensure the correct behavior of the network applications running on top of the distributed controllers. Such control traffic can be in-band or out-band.

However, distributing the network control state across the SDN control plane affects the performance objectives of the control applications. In fact, many state distribution trade-offs arise, as discussed by Levin et al. (2012), such as the trade-off between application state consistency/staleness (state synchronization overhead) and application performance (objective optimality), and the trade-off between application logic complexity and robustness to inconsistency.

More generally, Brewer's CAP theorem applied to networks (Panda et al. 2013) investigates the trade-offs involved between consistency (C), availability (A) and partition-tolerance (P). It states that, in SDN networks, it is generally only possible to achieve two out of the three desirable properties: CA, CP or AP.

However, in the context of modern and scalable distributed database systems (DDBSs), Abadi's PACELC theorem (Abadi 2012) is believed to be more relevant as it combines in a single complete formulation the CAP theorem trade-offs, and in the absence of partitions (E), the latency (L)/consistency (C) trade-off. Many popular modern DDBSs do not guarantee strong consistency by default, as stated by CAP. Conversely, they come with trade-offs that are better warranted/represented by the PACELC alternative. For example, Amazon's Dynamo (Sivasubramanian 2012), Facebook's Cassandra (Lakshman and Malik 2010) and Riak are PA/EL systems, MongoDB is PA/EC, Yahoo's PNUTS is PC/EL and finally BigTable and HBase are PC/EC systems.

In this context, we argue that SDN is bringing the network design much closer to the design of distributed database systems. In the same way, we argue that PACELC can apply to large-scale SDN controller platforms; in the same way, it applies to modern and scalable NoSQL DDBSs.

4.3.2. *Consistency models in SDN*

Many architectures have been proposed to support distributed SDN controllers, with the goal of improving the scalability, reliability and performance of SDNs. Two main consistency models are used by current controller platforms, as outlined below.

4.3.2.1. *The strong consistency model*

In SDN, the strong consistency model guarantees that all controller replicas in the cluster have access to the most updated network information at all times. This comes at the cost of increased state synchronization delay and communication overhead, especially in large-scale deployments. In fact, strong consistency relies on a blocking synchronization process that keeps the switches from reading the data – unless the controllers are fully updated – thereby affecting network availability and scalability.

Strong consistency is a requirement for certain applications that favor consistency and correctness properties over availability. In current controller platforms, strong consistency is usually achieved using Paxos, RAFT (Ongaro and Ousterhout 2014) and similar protocols.

4.3.2.2. *The eventual consistency model*

In SDN, the eventual consistency model takes a relaxed approach to consistency by assuming that all controller replicas will "eventually" converge and become consistent throughout the network. This means that controllers may temporarily present an inconsistent network view, allowing for some stale data to be read and potentially causing transient incorrect application behavior.

Many applications opt for eventual consistency to guarantee high availability and performance at scale. Modern DDBSs, including Dynamo (Sivasubramanian 2012) and Cassandra (Lakshman and Malik 2010), support eventual consistency settings by default in exchange for extremely high availability (fast data access) and scalability.

4.3.2.3. *Adaptive consistency models*

Recent research in SDN (Aslan and Matrawy 2016; Sakic et al. 2017; Aslan and Matrawy 2018; Sakic and Kellerer 2018; Sakic et al. 2018) has introduced the concept of adaptive consistency in the context of distributed SDN control. Unlike static consistency approaches, adaptively consistent controllers adjust their consistency level at runtime to reach the desired application performance and consistency requirements. This alternative offers many benefits; it spares application designers the task of developing complex applications that require the implementation of multiple consistency models, it provides applications with robustness against sudden network conditions, and it reduces the overhead of state distribution across controllers without compromising application performance (Aslan and Matrawy 2016).

In the community of modern database systems, the need for adaptable consistency where the consistency level is decided dynamically over time based on various factors has been recognized. Many adaptive consistency models have been proposed, such as QUOROM-based consistency (Kumar 2016), RedBlue consistency (Li et al. 2012), Chameleon and Harmony (Chihoub et al. 2013), delta consistency (Aslan and Matrawy 2016) and continuous consistency (Yu and Vahdat 2000). In addition, most modern database systems such as Cassandra (Lakshman and Malik 2010) and Dynamo (Sivasubramanian 2012) are currently equipped with an adaptive consistency feature, offering multiple consistency options with tunable parameters for application developers.

In our opinion, all of the above consistency models could be leveraged by the SDN community to build adaptively consistent SDN controllers.

4.4. Consistency models in ONOS

In this chapter, we are particularly interested in the open-source Java-based ONOS controller. In this section, we describe in detail the ONOS approach to state consistency in a distributed controller setting.

To achieve high availability, scale-out and performance, the ONOS controller platform supports a physically distributed cluster-based control plane architecture, in which each controller is responsible for handling the state of a subsection of the network. To maintain the logically centralized network view, local controller state information is disseminated across the cluster in the form of events that are shared by ONOS's distributed core. The latter consists of core subsystems tracking different types of network states being stored in distributed data structures and requiring different coordination strategies. Two main state consistency schemes are implemented in ONOS's subsystem stores to provide two different levels of state consistency: strong consistency and eventual consistency.

4.4.1. *Strong consistency in ONOS*

To ensure strong consistency among replicated network states, ONOS uses (since version 1.4) the Atomix framework that is based on the RAFT consensus protocol (Ongaro and Ousterhout 2014). For instance, the store for switch-to-controller mastership (mapping) management is handled in a strongly consistent manner using that framework. Moreover, ONOS features a set of core distributed state management primitives that can be leveraged by application developers to implement their application-specific stores. In this respect, applications whose state is maintained in a strongly consistent fashion can leverage the ConsistentMap distributed primitive, which guarantees strong consistency for a distributed key-value store.

4.4.2. *Eventual consistency in ONOS*

For eventually consistent behaviors, ONOS employs an optimistic replication technique complemented by a background gossip/anti-entropy protocol. For instance, the stores for devices, links, and hosts are managed in an eventually consistent manner. The distributed topology store is also eventually consistent since it relies on the distributed versions of the device, link and host stores. For the eventual consistency option, ONOS offers the EventuallyConsistentMap distributed primitive for control programs and applications. The latter can create different instances of these primitives for managing their eventually consistent application-specific states.

4.4.2.1. *Optimistic replication*

Optimistic replication is a key technology that is used in large-scale distributed data sharing systems, meeting the goal of achieving higher availability and scalability compared to strongly consistent systems. This strategy for replication propagates changes in the background and discovers conflicts after they occur. It is based on the "optimistic" assumption that inconsistencies rarely occur and that replicas will converge after some time, thus providing eventual consistency guarantees.

In ONOS, optimistic replication is used in the distributed maps. Whenever an update occurs in the store managed by one controller, the associated EventuallyConsistentMap replicates events immediately to the rest of the controllers. This means that maps on each controller will become closely in sync (apart from a small propagation delay) in cases where the controllers are functioning properly. On each controller, updates are added to an EventAccumulator as they are written.

4.4.2.2. *Gossip-based anti-entropy*

Controllers that rely purely on optimistic replication may become progressively out of sync, especially in the event of node failures and partitions or when updates are missed or dropped. The anti-entropy protocol ensures that replicas are back in sync by resolving discrepancies and the entire cluster converges fairly quickly to the same state.

In ONOS, the gossip-based anti-entropy mechanism is a lightweight peer-to-peer background process that runs periodically. At fixed intervals (3–5 s), each controller randomly chooses another controller and they both exchange information in order to compare the actual content (entries) of their distributed stores (based on timestamps). After synchronizing their respective topology views, the controllers become mutually consistent. This reconciliation approach proves useful in fixing controllers when their state drifts slightly and quickly synchronizing a newly joining controller with the rest of the controllers in the cluster.

4.5. The proposed adaptive consistency for ONOS

In this section we explain our approach, which is mainly aimed at optimizing consistency management in ONOS.

4.5.1. *A continuous consistency model for ONOS*

As explained in section 4.3.2.3, an adaptive consistency model offers many benefits for the distributed SDN controllers. In particular, the ONOS controllers can benefit from the continuous consistency model proposed in Yu and Vahdat (2000). The latter is based on a middle-ware framework (called TACT) for adaptively tuning the consistency and availability requirements for replicated online services, following the continuous consistency concept. In contrast to the strong consistency model (which imposes performance overheads and limits availability) and optimistic consistency models (which provide no bounds on system inconsistencies), the continuous consistency model explores the semantic space between these two types of traditional models. It offers a continuum of intermediate consistency models (*multi-level consistency*) with tunable parameters. These quantifiable degrees of consistency can be exploited by applications to explore, at runtime, their own trade-offs between consistency and availability, while taking into account the changing network and service conditions. More specifically, TACT bounds the amount of inconsistency and divergence among the replicas in an application-specific manner. Basically, applications specify their consistency semantics through *conits*, a set of metrics that capture the consistency spectrum: *numerical error*, *order error*, *staleness*.

Hence, for each conit, consistency is quantified continuously along a three-dimensional vector:

$$Consistency = (NumericalError, OrderError, Staleness) \qquad [4.1]$$

Numerical error bounds the discrepancy between the value delivered to the application client and the most consistent "final" value. *Order error* bounds inconsistency by the number of tentative/unseen writes at any replica. *Staleness* places a real-time bound on the delay for propagating the writes among the replicas.

In our opinion, many features from the discussed continuous consistency spectrum can indeed be incorporated when rethinking the ONOS strategy to state consistency, especially in the context of large-scale deployments.

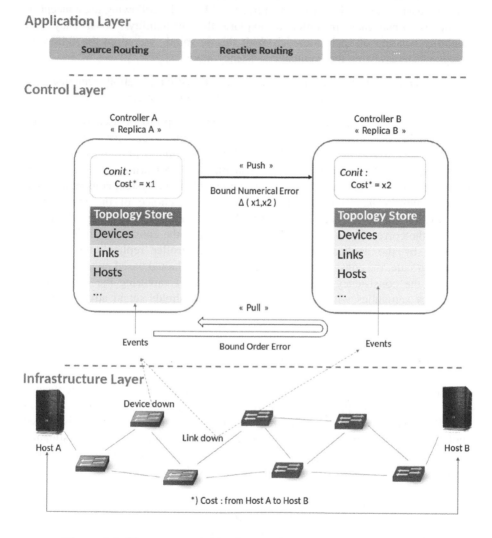

Figure 4.1. *The proposed adaptive consistency strategy. For a color version of this figure, see www.iste.co.uk/bannour/software.zip*

4.5.2. *Our consistency adaptation strategy for ONOS*

We propose to maintain the strong consistency model implemented in ONOS for applications requiring strict consistency guarantees, but we suggest turning the eventual consistency model into an adaptive tunable model following the concept of continuous consistency in order to explore the availability, consistency and scalability benefits of such a model. With this in mind, we adopt the following strategy when reviewing the eventual consistency model:

– we keep the current implementation of the optimistic replication technique used for replicating events and updates among controllers in the cluster;

– we bring significant changes to the anti-entropy protocol used for eventual consistency in ONOS. Instead of running the anti-entropy process for each controller replica periodically at fixed intervals (each 3–5 s) (*voluntary anti-entropy*) causing too much overhead and affecting system scalability and performance, we argue that the anti-entropy process should only be scheduled when the system consistency is at risk (*compulsory anti-entropy*). In other words, the choice of the anti-entropy reconciliation period for each controller replica (*per-replica consistency*) should be based on the correctness of the system with respect to the consistency requirements expressed by the applications. Thus, at each controller replica and for each application state, the consistency level is dynamically adapted based on the computed values of the consistency metrics (see equation [4.1]) capturing the application's consistency semantics with respect to the given thresholds set in advance by the application.

4.5.3. *Our implementation approach*

To implement our state consistency adaptation approach, we consider a replicated source routing SDN application running on top of a cluster of multiple ONOS controllers. The cost-based source routing application operates on a distributed topology graph for computing the shortest-path costs between source and destination hosts. Since the topology graph state is handled in an eventually consistent manner, the application's state is also considered to be eventually consistent.

In our routing application f, the path between source host A and destination host B (see Figure 5.1) may be defined as a *conit*. Moreover, we argue that an important consistency requirement for our control application is the result optimality of the instant path computation cost (in terms of hop-count in our case) which is captured by the *numerical error* metric. The numerical error of our conit C can be defined as the relative difference between the value of the "shortest-path" cost x_{local} as perceived by a local replica and its final "optimal" value $x_{optimal}$ at a replica that has reached some "final" consistent state. This error is continuously bounded at runtime using an application-defined threshold $T(f)$ (a percentage) as follows:

$$NumericalError(C_f) = (\frac{|x_{local} - x_{optimal}|}{x_{optimal}}) < T(f) \qquad [4.2]$$

Finally, it is worth noting that other consistency semantics for the source routing application might be expressed using *staleness* and *order error*.

To implement our approach, we introduce some modifications to the ONOS Java source code. We start by developing our adaptive source routing application (similar to the Intent Forwarding Application) for computing the shortest-path cost between host A and host B. Running on each ONOS instance, our application obtains information from the in-memory topology cache maintained by each ONOS instance (DistributedTopologyStore). Whenever an update occurs in the topology graph (e.g. links/devices failing or joining), our application detects that topology change and updates the "local" shortest-path cost between hosts A and B accordingly. We also modify the implementation of the EventuallyConsistentMap distributed primitive, especially for the eventually consistent stores that feed the topology store (e.g. link and device stores). Indeed, we focus on the BackgroundExecutor service of the eventually consistent maps, which runs the background anti-entropy tasks, and we propose a new implementation of the Runnable interface used for executing the scheduled anti-entropy thread.

In fact, instead of sending the anti-entropy advertisement messages periodically every 3–5 s between the controllers, as is the case in ONOS, we propose to run, at each replica, a periodic check on the consistency of other replicas with respect to the application's shortest-path cost state by computing at runtime the relative *numerical error* defined in equation [4.2]. In the event of a controller failure, the rest of the controllers in the cluster that detect that failure, keep a screenshot of their own topology graph at the moment of the failure. During the periodic consistency check, they use that stored topology graph to estimate the inconsistency of the failed controller, which is equal to the relative difference between the "local" shortest-path cost (computed based on the current topology graph state) and the shortest-path cost, as perceived by the controller after recovery (computed based on the *stale* topology graph state being stored).

When the failed controller recovers, the rest of the controllers make an anti-entropy decision based on the checked numerical error. If the error exceeds an "alarming" consistency threshold set in advance by the application, then an anti-entropy process is launched to fix the failed controller state. This is achieved by synchronizing the controllers' eventually consistent stores that feed the topology view. In the opposite case, the inconsistency is considered tolerated by the application, and an anti-entropy session might be scheduled afterwards in case the controller state significantly drifts away.

4.6. Performance evaluation

4.6.1. *Experimental setup*

Our experiments are performed on an Ubuntu 16.04 LTS server using ONOS 1.13. We use the network emulator Mininet 2.2.1, which can create virtual switches and hosts, and can connect to the ONOS controllers. We also use an ONOS-provided script (*onos.py*) to start an emulated ONOS network on a single machine, including a logically centralized ONOS cluster, a modeled control network and a data network. Wireshark is used as a sniffer to capture the inter-controller traffic, which uses TCP port 9876.

To validate our proposed approach on ONOS, which we will refer to as ONOS-WAC (ONOS-With Adaptive Consistency), we have considered many test scenarios. In each scenario, we run a cluster of N ONOS controller instances, controlling a Mininet network topology of S switches (see Table 4.1).

Test scenarios	N Controllers	S Switches	F Controller Failure scenarios
No. 1	3	16	2
No. 2	5	36	3
No. 3	7	64	4
No. 4	9	100	4
No. 5	10	121	5

Table 4.1. *Test scenarios*

In order to create state inconsistencies among the controller instances in the cluster with respect to the shortest-path cost state of our source routing application, we create different controller failure scenarios F. Shortly after a controller failure (in F) in a specific scenario S_i, we consider changing the network topology by taking down network switches and links along the shortest-path (computed by the application) between source host A and destination host B. Thus, after recovery, the controller will have an inconsistent network topology view compared to the rest of the controllers in the cluster. According to our proposed approach, inconsistency in the network topology view affects the optimality of the shortest-path cost computation performed by the source routing application instance running on top of the recovered controller. The induced *numerical error* is likely to trigger a synchronization process achieved by anti-entropy tasks (see section 4.5.3).

4.6.2. *Results*

In Scenario no. 1 (scenario with three controllers, as shown in Table 4.1), we adopt the methodology described in section 4.6.1. In Figure 4.2, we show the inter-controller traffic captured during the test scenario period in an ONOS cluster (Figure 4.2(a)) and in an ONOS-WAC cluster (Figure 4.2(b)). After running the Mininet topology according to Scenario no. 1, the same event sequence is performed for both ONOS and ONOS-WAC clusters. For instance, the first traffic peak in both figures (at $t = 90\,$s) corresponds to a "pingall" Mininet CLI command executed for topology discovery. At $t = 150\,$s, we simulate a failure scenario by taking down one controller instance. This action is followed by other topology changes (e.g. links down) corresponding to the subsequent peaks in both figures. At $t = 180\,$s, we bring back the failed controller which results in a traffic peak that appears to be more significant in the case of ONOS-WAC. This increase in traffic is due to the anti-entropy process, which has been triggered by an inconsistency (*numerical error*) value that exceeded the application threshold. Conversely, in the ONOS network, the anti-entropy traffic is generated periodically over the test period regardless of the observed inconsistencies. Likewise, at $t = 280\,$s, we repeat the same scenario following the same event sequence but considering the failure of a different controller.

In the test scenario described above, the application inconsistency threshold was set to 0%, triggering the start of anti-entropy sessions for any observed inconsistencies in the considered application state. In the following test experiments, we repeat the same scenario (Scenario no. 1), but we consider varying the source routing application's inconsistency threshold. As shown in Figure 4.3, ONOS shows an average inter-controller overhead equal to 315 kbps regardless of the application inconsistency threshold. On the other hand, ONOS-WAC shows relatively less inter-controller overhead (due to low anti-entropy overhead). The latter is impacted by the application's consistency requirements. For example, in the case of strict consistency requirements (application threshold between 0% and 30%), inconsistencies occurring in the application state are more likely to trigger the anti-entropy reconciliation sessions causing much more overhead, when compared to the case of less strict consistency requirements (application threshold between 40% and 50%).

In order to assess the gain in anti-entropy overhead of our adaptive consistency model implemented on ONOS, compared to the eventual consistency model of ONOS, we consider estimating the rate ($R(S_i)$) of increase in anti-entropy overhead of ONOS with respect to ONOS-WAC (see equation [4.3]) as a function of the number of controllers in the cluster (following Scenarios no. 1, 2, 3, 4, 5) (see Figure 4.4).

$$R_i(S_i) = 1 - [\frac{A(S_i) - B(S_i)}{C(S_i) - B(S_i)}] \qquad\qquad [4.3]$$

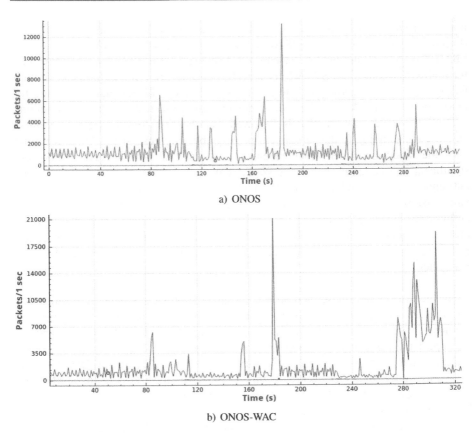

Figure 4.2. *Scenario no. 1: captured inter-controller traffic (in packets per second) during the test scenario period (using Wireshark)*

$-A(S_i)$: the inter-controller overhead generated by ONOS-WAC *after* the event sequence (section 4.6.2) following Scenario S_i;

$-B(S_i)$: the inter-controller overhead generated by ONOS-WAC *before* the event sequence (the overall inter-controller traffic without the anti-entropy traffic);

$-C(S_i)$: the inter-controller overhead generated by ONOS *after* the event sequence (section 4.6.2) following Scenario S_i.

As shown in Figure 4.4, the gain in anti-entropy overhead when adopting ONOS-WAC grows almost linearly with the number of controllers in the cluster. For example, in Scenario no. 5 (corresponding to 10 controllers in the network cluster), the gain in anti-entropy overhead has reached 25%.

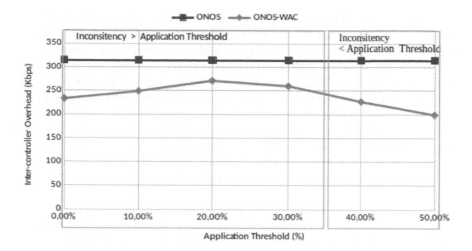

Figure 4.3. *Scenario no. 1: inter-controller overhead in ONOS and ONOS-WAC according to the application threshold*

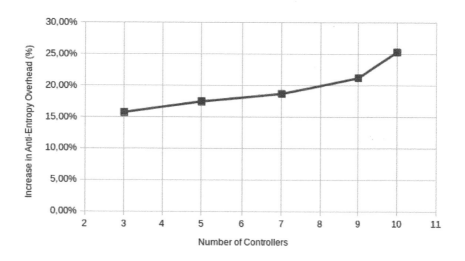

Figure 4.4. *Gain in anti-entropy overhead of ONOS-WAC with respect to ONOS according to the number of controllers in the cluster. For a color version of this figure, see www.iste.co.uk/bannour/software.zip*

4.7. Conclusion

In this chapter, we have investigated the use of adaptive consistency for distributed ONOS controllers. Our approach turned ONOS's eventual consistency model into an adaptable multi-level consistency model following the concept of continuous consistency. The latter delivers the performance and availability benefits of an eventual consistency model but has the additional advantage of controlling the state inconsistencies in an application-specific manner. Our consistency adaptation strategy was implemented for a source routing application on top of ONOS. Moreover, ensuring the application's state consistency requirements (specified in the given SLA), our results showed a substantial reduction in the anti-entropy reconciliation overhead, especially in the context of large-scale networks. As future work, we consider extending our adaptive consistency approach to the optimistic replication technique used in ONOS's eventual consistency model by leveraging multiple replication degrees as well as the geo-placement of the controller replicas.

Although the main focus of this work is on dynamically adjusting the consistency level of application states (which use controller states), we plan to extend our approach to the controller states (internal controller applications). Indeed, the long-term goal of this work is to design adaptively consistent controllers that adjust both control and application plane consistency levels under changing network conditions.

5

Adaptive and Continuous Consistency for Distributed SDN Controllers: Quorum-Based Replication

5.1. Introduction

Existing SDN controller platforms have been created according to different SDN control plane designs with the aim of meeting specific requirements in terms of scalability, high availability and performance. Consistency has also been regarded as an essential design principle for the distributed SDN controller platforms. The latter use conventional consistency models to manage the distributed state among the controllers in the cluster. As explained in section 4.3.2, the consistency models used in SDN can be categorized into *strong*, *eventual* and *weak* (OpenDayLight; ONOS). These static and standard consistency models (OpenDayLight; ONOS) have both advantages and drawbacks.

In large-scale SDNs, the *strong consistency* control model might be extremely expensive and costly to maintain for certain applications. Indeed, it requires important synchronization efforts among the controller replicas at the cost of causing serious network scalability and performance issues. In contrast, the *eventual consistency* control model implies less inter-controller communication overhead as it sacrifices the strict consistency guarantees for higher availability and improved performance. In practice, many scalable control applications running in modern distributed storage systems such as Apache's Cassandra (Lakshman and Malik 2010) and Amazon's DynamoDB (Sivasubramanian 2012) opt for eventual consistency to provide such requirements on a large scale. However, these applications might suffer from the associated relaxed (weak) consistency guarantees that may temporarily allow for too much inconsistency.

Recent research works in the area of distributed SDN control have explored the concepts of *adaptive consistency* control for various applications (Aslan and Matrawy 2016; Sakic et al. 2017; Aslan and Matrawy 2018; Sakic and Kellerer 2018; Bannour et al. 2018a). Such categories of consistency models follow different adaptation strategies which mainly focus on dynamically adjusting the levels of consistency at runtime under various network conditions, in order to meet the application-defined consistency and performance needs.

Unlike strong and eventual consistency options, adaptive consistency control models leverage the broad space of intermediate consistency degrees between these two extremes. Indeed, they use time-varying consistency levels to support balanced real-time trade-offs between the desired consistency and performance requirements, which can be specified in the application-defined service-level agreements (SLAs) (Terry et al. 2013).

In this chapter, we propose an adaptive consistency model (based on eventual consistency) for the ONOS controller applications that are deployed in large-scale networks (Bannour et al. 2019). Most notably, we target the class of applications that tolerate relaxed forms and degrees of eventual multi-consistency for the sake of scalability and performance, but yet can benefit from improved consistency features.

More specifically, our state consistency adaptation approach was implemented for a CDN-like application we developed on top of the distributed open-source ONOS controllers. It mainly consists of changing ONOS's eventual consistency model to an adaptive consistency model by turning ONOS's optimistic replication technique into a more scalable replication strategy following Quorum-replicated consistency models. Indeed, the adaptive consistency strategy we propose in this chapter focuses on improving ONOS's replication mechanism. It uses the *read and write Quorum parameters* as adjustable *control knobs* for a fine-grained consistency tuning rather than relying on anti-entropy reconciliation mechanisms (see Chapter 4) (Bannour et al. 2018a). The main objective is to find optimal Quorum replication configurations at runtime that, under changing network conditions and varying application workloads, achieve balanced trade-offs between the application's continuous performance (*latency*) and consistency (*staleness*) requirements. These real-time trade-offs should provide minimal application inter-controller overhead while satisfying the application-defined thresholds specified in the given application SLA.

This chapter is organized as follows. In section 5.2 we conduct a background review of eventual consistency models in modern distributed data store systems. Inspired by the modern consistency techniques used in these scalable data stores, in section 5.3 we present our adaptive and continuous Quorum-based consistency model for the distributed ONOS controllers in large-scale deployments. In section 5.4, we describe our methodology for implementing the proposed consistency strategy on a CDN-like application that we designed on top of the ONOS controllers (Bannour et al. 2020). Finally, section 5.5 elaborates on the test scenarios we developed to evaluate our proposal and discusses the experimental results.

5.2. Background on eventual consistency in distributed data stores

5.2.1. *Consistency and performance metrics*

Guaranteeing the consistency of replicated data in distributed database systems has always been a challenging task. Today's fundamental consistency models (e.g. strong consistency, sequential consistency, causal consistency, eventual consistency) ensure different discrete levels and degrees of consistency guarantees. For instance, the strong consistency model provides up-to-date data but at the cost of high latency and low throughput. As a result, weaker forms of consistency (in the consistency spectrum), most notably the popular notion of eventual consistency, have been widely adopted in modern distributed data stores, which need to be highly available, fast and scalable (Sivasubramanian 2012; Lakshman and Malik 2010). However, despite often being acceptable and desirable in practice for the latency and throughput benefits they offer, eventual consistency models provide no bounds on the inconsistency of data they return. Another major limitation of these models is that the trade-offs they make between consistency and performance (latency) are difficult to evaluate. In fact, measuring the concrete consistency guarantees of eventually consistent distributed stores remains challenging.

Yu and Vahdat (2000) proposed the TACT framework, which fills in the consistency spectrum by providing a continuous conit-based and multi-dimensional consistency model. The latter can be leveraged by replicated Internet services to dynamically choose their own tunable and fine-grained consistency-performance and consistency-availability trade-offs based on client, service and network characteristics. In TACT, the authors quantify consistency by bounding the amount of divergence of the replicated data items in an application-specific manner using three application-independent metrics: *numerical error*, *order error* and *staleness*. Moreover, Bailis et al. (2012, 2014) presented an approach based on a set of probabilistic models to predict the expected consistency guarantees as measured by the *staleness* of reads observed by client applications in eventually consistent DynamoDB-style partial Quorum systems. The authors introduced the WARS probabilistically bounded staleness (PBS) model, which provides bounds on the expected staleness in terms of both versions (using the *k-staleness* metric) and wall clock time (using the *t-visibility* metric). Another interesting work found in Chihoub et al. (2012) proposes an automated self-adaptive consistency approach called Harmony, which embraces an intelligent estimation of the *stale read rate* metric in Cloud storage systems, making it possible to automatically adjust the consistency level at runtime according to application needs. This was achieved by elastically scaling up or down the number of replicas involved in read operations to preserve a low tolerable fraction of stale reads. When compared to the static eventual consistency approach in Cassandra, Harmony significantly enhances the consistency guarantees by reducing the rate of stale reads while adding only minimal latency. Moreover, when compared to the strong consistency model in Cassandra, Harmony improves the performance of the system by increasing the overall throughput while maintaining the desired consistency requirements of the applications.

5.2.2. *Adaptive consistency control*

Modern distributed database systems supporting standard eventual consistency models suffer from the inevitable trade-offs between consistency, availability and request latency. To overcome this major limitation, these storage systems have introduced the concept of adaptive consistency in order to find appropriate consistency options depending on application requirements and system conditions. In the literature, adaptive consistency techniques have been broadly classified into two categories: *user-defined* and *system-defined* (Kumar 2016).

In contrast to user-defined adaptive consistency methods, in which data and operations need to be mapped in advance to the desired consistency levels (using specific parameters), system-defined adaptive consistency methods take into account the fact that user and system behaviors might change dynamically over time, making the consistency decision-making process challenging for application developers. This is why system-defined techniques usually rely on system intelligence and adaptability to automatically provide fine-grained control over the consistency guarantees at runtime. Accordingly, many factors can be considered to dynamically estimate and predict the appropriate system consistency, including data access patterns, system load and the application's consistency SLAs, as discussed in section 5.2.1. One well-known form of system-defined adaptive consistency is the continuous consistency model used in TACT (Yu and Vahdat 2000).

Additionally, it is worth mentioning that designing system-defined adaptive consistency (falling within the scope of this book) requires careful consideration of the appropriate consistency adaptation strategy. In particular, existing adaptive mechanisms use different control knobs to be configured for consistency tuning such as the *consistency level*, the *artificial read delay*, the *replication factor* and the *read repair chance* (Nguyen Ba 2015).

5.2.3. *Existing modern tunable consistency systems*

To ensure eventual consistency in existing distributed data store systems, different replication mechanisms and reconciliation techniques can be implemented. The most commonly used replication mechanism is optimistic (lazy) replication, which is believed to offer high availability, performance and scalability. There are several variants of optimistic replication systems (Bouajjani et al. 2014), but the common basic concept is to passively replicate the updates to other replica nodes in the system and let them be read by clients without the need to wait for a prior synchronization of all the copies. In some implementations, a minimum number of nodes (called a Quorum) are involved in the updates. Other design choices, like the number of masters in the system, might also result in different optimistic replication system variants.

Along with optimistic replication, eventual consistency systems may resort to extra conflict resolution and reconciliation techniques in order to reconcile differences (after they occur) between multiple copies of distributed data. The most appropriate approach to reconciliation depends on the considered application. For example, Amazon's DynamoDB (Sivasubramanian 2012) uses vector clocks for conflict resolution. In general, reconciliation strategies include read repair, anti-entropy recovery, write repair and asynchronous repair operation mechanisms.

Popular distributed (Cloud) storage systems, most notably Apache's Cassandra (Lakshman and Malik 2010), Amazon's DynamoDB (Sivasubramanian 2012), Riak (Klophaus 2010) and Project Voldemort "by default" opt for eventual consistency guarantees in exchange for extremely high availability. However, these systems attempt to provide the applications with some control over the consistency and performance trade-offs via built-in settings and features. Indeed, they extend the concept of eventual consistency by offering tunable consistency levels for application developers and users based on DynamoDB-style Quorum replication policies.

In Cassandra, the consistency level specifies the size of a Quorum for reads and writes, which is the appropriate number of replicas in the cluster that must acknowledge a read or write operation before considering the operation successful. The native and well-known consistency options (levels) in Cassandra are: ONE replica, a QUORUM of replicas and ALL of the replicas. Accordingly, different choices of read and write consistency levels (Quorums) ensure different consistency guarantees. For instance, to achieve the highest strong consistency, different Quorum configurations may be selected, but they must satisfy the overlapping Quorum property between read and write replica sets (*strict Quorums*). On the other hand, to provide acceptable consistency with improved availability (minimum latency), it is desirable to use weaker forms of consistency such as the default eventual consistency option. Such weak consistency levels can be achieved through different Quorum configurations that do not satisfy the overlapping Quorum intersection property (*partial (non-strict) Quorums*).

As a result, modern storage systems like Cassandra can be classified under the category of *user-defined* adaptive consistency – as discussed in section 5.2.2 – given that they offer multiple consistency options. However, although these systems offer adaptive consistency on top of tunable consistency models, aimed at creating balanced trade-offs between consistency and performance, it is usually difficult for application developers to decide on the required consistency options for a particular request in advance (Kumar 2016).

5.3. The proposed adaptive Quorum-inspired consistency for ONOS

In this chapter, we propose a novel Quorum-based and *system-defined* adaptive consistency model for the distributed ONOS controllers. Our approach was partly inspired by the Quorum-replicated consistency techniques used by the modern data store systems discussed in section 5.2.3.

The ONOS approach to state consistency in the latest releases is described in detail in section 4.4. It mainly relies on two consistency schemes that provide two levels of consistency: strong consistency and eventual consistency. While the strong consistency model is leveraged by ONOS controller applications that require strong consistency and correctness guarantees, the eventual consistency model is intended for ONOS controller applications that favor scalability and performance over strict consistency.

In this chapter, we target the second class of scalable control applications that have optimistic relaxed consistency needs but that can benefit from improved performance and automated SLA-aware consistency tuning at scale, as offered by our adaptive continuous consistency strategy.

5.3.1. *A continuous consistency model for ONOS*

As explained in section 4.5.1, the applications on top of the ONOS controllers can benefit from the continuous consistency model introduced with TACT (Yu and Vahdat 2000) by continuously and dynamically specifying their consistency requirements using three application-independent metrics to capture the consistency spectrum and bound consistency: *numerical error*, *order error* and *staleness*.

Here, we focus on applications whose application-specific consistency semantics can be expressed using the staleness of data as a metric to quantify the level of consistency. With such SLA-style consistency metrics, these applications can prevent the challenges related to potentially *unbounded staleness* as in eventual consistency.

Generally, the staleness metric measures data freshness in distributed data stores; it describes how far a given replica lags behind in data operations in comparison to up-to-date replicas, either expressed in terms of time or versions. In the literature, the notion of data staleness falls indeed into two common categories: staleness in time (*time-based staleness*) (Yu and Vahdat 2000; Bailis et al. 2012) and staleness in data version (*version-based staleness*) (Bailis et al. 2012).

In TACT (Yu and Vahdat 2000), the staleness metric places a real-time bound on the amount of time before a replica is guaranteed to see a write accepted by a remote replica. In Bailis et al. (2014), the authors propose a probabilistic consistency framework that provides expected bounds on data staleness with respect to both versions and wall clock time in eventually consistent data stores. In their model, time-based staleness (*t_visibility*) describes the probability that a read operation, starting t seconds after a write commits, will observe the latest value of a data item (Bailis et al. 2012). On the other hand, version-based staleness (*k_staleness*) describes how many versions the value returned by a read lags behind the most recent write. It is measured as the probability of returning a value within a bounded number k of versions.

We adopt the data staleness metric from a strictly time-based perspective. In our SDN controller application, we characterize staleness by an "Age of Information (AoI)" timeliness metric (Zhong et al. 2018) which describes the difference between the query time of a data item and the last update time on that item. If the last successfully received update was generated at time $u(t)$, then its age at time t is $\Delta(t) = t - u(t)$.

Applications on top of the distributed ONOS controllers could also benefit from SLA-style performance requirements to continuously specify their own fine-grained trade-offs between performance and consistency. We consider the read request latency/delay as our performance metric. In addition, we evaluate the inter-controller communication overhead for our ONOS application.

More detailed information about the way we measure our continuous consistency and performance metrics when implementing our state consistency approach for the new controller application that we designed on top of ONOS is provided in section 5.5.1.

5.3.2. *Our Quorum-inspired consistency adaptation strategy for ONOS*

5.3.2.1. *Quorum consistency*

As explained in section 5.2.3, Quorum-replicated systems ensure different consistency guarantees.

– Strong consistency can be guaranteed with *strict Quorums* which satisfy the condition that sets of replicas written to and read from need to overlap:

$R + W > N$, given N replicas and read and write Quorum sizes R and W.

– Eventual consistency occurs with *partial Quorums* which fulfill the condition that sets of replicas written to and read from need not overlap:

$R + W \leq N$, given N replicas and read and write Quorum sizes R and W.

Traditionally, partial Quorum-replicated systems ensure eventually consistent guarantees, with no limit to the inconsistency of the data returned, which may not be acceptable for certain applications. However, with the PBS model (Bailis et al. 2012), it has been possible for applications to analyze the staleness of the data returned, quantify the consistency level, and therefore measure the latency-consistency trade-offs for partial Quorum systems.

Building on these concepts, we propose an adaptive consistency model for the ONOS applications using partial Quorums, given the latency and scalability benefits they offer. To measure the consistency semantics (e.g the staleness metric) of these applications, and thus meet their consistency requirements (e.g bounded staleness), we leverage the continuous consistency model discussed in section 5.3.1.

Furthermore, using eventually consistent partial Quorums, it is possible to configure the size of read and write Quorums, denoted respectively as R and W such that $R + W \leq N$, to ensure various consistency levels (e.g degrees of staleness). These multiple Quorum configurations enable the applications to achieve different consistency-latency trade-offs.

5.3.2.2. *Adaptive architecture*

We propose to turn the eventual consistency model into an adaptive and continuous tunable consistency model using partial Quorums. The proposed model uses the Quorum replication parameters as the control knob, allowing for an adaptive fine-grained tuning and control over the consistency-performance trade-offs. In the following, we describe the main architecture components of our adaptive consistency model.

5.3.2.2.1. Automatic Module

The choice of the size of read and write Quorums used when executing read and write operations is a fundamental factor that affects the application's consistency guarantees but also the performance provided by the network system. However, selecting the right Quorum configuration is a non-trivial task. Our Automatic Module attempts to find the optimal configuration of the read and write Quorum sizes while taking into account the current application workload conditions. The main objective is to minimize the overhead generated by the application (scalability challenge), and potentially other network and application metrics, while satisfying the consistency and performance SLAs specified by the application.

This module is fed with a set of application workload characteristics that are gathered by the Workload Identifier Module. In our case, it relies on a machine learning module to predict the expected optimal configuration of the Quorum parameters for the determined workload, and then feed them to the reconfiguration module.

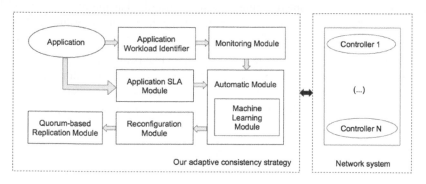

Figure 5.1. *Architectural overview of our adaptive Quorum-based consistency strategy*

5.3.2.2.2. Machine Learning Module

This module uses reinforcement learning (RL); an area of machine learning (ML) inspired by behaviorist psychology, and concerned with how software agents take actions in an environment so as to maximize some notion of cumulative reward.

More specifically, we use a Q-learning (QL) model-free RL technique (Mellouk et al. 2013). The main idea is to train an *agent* which interacts with its *environment* by performing *actions* that change the environment, going from one *state* to another. These actions result in a *reward* received by the agent as an evaluation of its actions (reinforcement) (see Figure 5.2). In this way, the agent learns certain rules and develops a strategy, referred to as a *policy*, for choosing actions that maximize its reward.

The QL update rule makes use of the so-called action-value function, commonly known as the *Q-function*, representing the "quality" of a certain action in a given state. The expression of the Q-function is given by the following Bellman equation:

$$Q(s_t, a_t) \leftarrow (1 - \alpha) \times \underbrace{Q(s_t, a_t)}_{\text{old value}} + \underbrace{\alpha}_{\text{learning rate}}$$

$$\times \left[\underbrace{r_t}_{\text{reward}} + \underbrace{\gamma}_{\text{discount factor}} \times \underbrace{\overbrace{\max_a Q(s_{t+1}, a)}^{\text{learned value}}}_{\text{estimate of optimal future value}} \right] \qquad [5.1]$$

The above Q-function is used to update the *Q-table* with Q-values at each *episode*. A Q-value is assigned to a possible pair of a state s_t and a current action a_t. The Q-function takes as input the pair (s_t, a_t), observes a new state s_{t+1} and returns the expected rewards of that action at that state. More specifically, the Q-function maps state-action pairs to the highest combination of the immediate reward r_t for that action with all discounted (using γ) future rewards that might be collected by later actions. The future rewards are computed using the maximum value of Q, given by $\max\limits_{a} Q(s_{t+1}, a)$, for all possible actions in the next state, assuming the agent continues to follow the optimal policy. We also note that the discount factor $\gamma \in [0; 1[$ determines the importance of future rewards with respect to immediate/current rewards, whereas the learning rate $\alpha \in]0; 1]$ determines to what extent newly acquired information (during the learning process) overrides the previous old information.

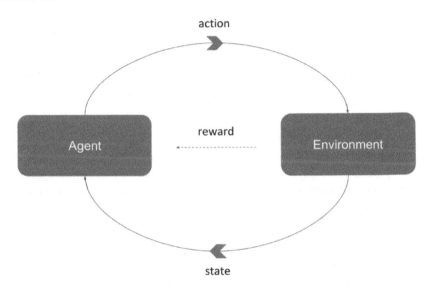

Figure 5.2. *Reinforcement learning (RL) architecture. For a color version of this figure, see www.iste.co.uk/bannour/software.zip*

We also note that the learning agent should achieve a good strategy for balancing the trade-off between exploration and exploitation, which is inherent to reinforcement learning. This dilemma consists of choosing the appropriate action at a given episode: either to *exploit* the environment by selecting the best action at that specific time step, given the current knowledge provided by the Q-table, or to *explore* the environment by choosing random actions. After each action, the agent is expected to update the Q-table.

In our case, the QL agent attempts to learn online the best combination of the read and write Quorum size parameters, respectively R and W, in an environment built using our Monitoring Module. An action is defined as an update of R and W to certain possible values, thereby transforming the environment to a state defined by a new estimation of the network (inter-controller overhead) and application (latency and staleness) metrics. In our case, one of four possible actions is allowed at each episode (i.e. incrementing R by one, decrementing R by one, incrementing W by one or decrementing W by one).

The reward received by the agent for updating the Quorum parameter values is a function of the read and write overheads to be minimized. The agent should also learn how to respect some constraints in order to satisfy the application requirements specified in the given SLA.

5.3.2.2.3. Reconfiguration Module

This module is able to dynamically adjust the values of the read and write Quorum sizes, denoted respectively as R and W. It essentially relies on the Automatic Module to optimize the configuration of the Quorum system. The reconfiguration process launched by this module is a non-blocking process that is able to re-configure at runtime the Quorum settings selected by the Automatic Module.

A more detailed description of the way the reconfiguration module sets the values of R and W at runtime is provided in section 5.5.2.1.

5.3.2.2.4. Quorum-based Replication Module

Given the Quorum replication settings, we adopt the following consistency strategy when reviewing the two main techniques employed by ONOS's eventual consistency model.

– Replication strategy: as explained in section 4.4.2.1, ONOS's eventually consistent stores employ an optimistic replication technique that consists of replicating local updates across all the controllers in the cluster, hence causing control plane overhead. Instead, we put forward a partial Quorum replication strategy, whereby an eventually consistent data store writes a data item on the local replica first and then sends it potentially to another set of replicas, obeying the given write Quorum parameter (W). On the other hand, to serve read requests, we propose that the eventually consistent data store fetches the data from the local replica first and then potentially from another set of replicas, depending on the given read Quorum (R). This is in contrast to ONOS's strategy in which the read requests are always processed by the local replica.

– Anti-entropy reconciliation mechanism: as explained in section 4.4.2.2, ONOS's optimistic replication strategy is complemented by a background anti-entropy mechanism. This periodic reconciliation approach ensures that the system state across all replicas eventually converges to the consistent state. This is particularly useful in repairing out-of-date replicas and fixing state inconsistencies, potentially resulting from controller failures. We assume the system is reliable as we experiment with well-functioning emulated network topologies in the absence of controller failure scenarios. Therefore, we propose to deactivate the anti-entropy protocol, and focus on ONOS's replication strategy. However, it is worth noting that using additional anti-entropy (*expanding partial Quorums* (Bailis et al. 2012)) might be useful in particular cases where state inconsistencies become high and can no longer be tolerated by the concerned applications.

5.3.2.2.5. Application SLA Module

This module offers the possibility for applications on top of ONOS to express their high-level SLA-style consistency and performance requirements, such as the staleness and latency guarantees. Accordingly, for a given ONOS application that we develop on top of ONOS, our consistency model continuously measures the real-time metrics involved in quantifying the consistency-latency trade-off. The Automatic Module translates these requirements into appropriate time-varying partial Quorum replication configurations (R, W, N) that achieve balanced trade-offs between the specified guarantees.

5.3.2.2.6. Workload Identifier Module

This module identifies the application's workload characteristics. It considers three different workloads that are representative of three different application scenarios (Couceiro et al. 2015). The first workload describes a read-intensive scenario where 70% of operations are read accesses. The second workload has a balanced ratio between read and write operations. Finally, the third workload represents a write-dominated scenario in which 70% of the generated operations are write accesses.

5.3.2.2.7. Monitoring Module

This module is responsible for periodically gathering the application traffic information in a non-intrusive manner. More specifically, the module measures the system KPIs (key performance indicators), for different read and write Quorum configurations and according to different application workload scenarios. These KPIs include the performance (e.g. response time) and consistency (e.g staleness) metrics related to client requests for specific application contents, as well as the generated read and write application overheads. These measurements are used by our Automatic Module (more particularly the Machine Learning Module) to learn the appropriate Quorum configurations online.

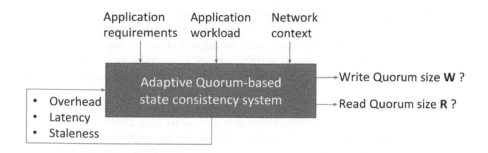

Figure 5.3. *The proposed adaptive consistency system*

5.4. Implementation approach on ONOS

In this section, we describe the implementation details for realizing the proposed consistency strategy on the Java-based open source ONOS controller platform. This strategy is explained in detail in the previous section and summarized in Figure 5.3.

5.4.1. *Design of a CDN-like application*

To validate our adaptive consistency approach, we developed a new distributed content delivery network (CDN) application running on top of a cluster of multiple ONOS controllers in an emulated SDN network. Our application replicates contents from content providers to hosting cache servers that are located in multiple geographical locations (ONOS domains) close to users. These cache servers are Mininet hosts that run simple HTTP web servers. We propose to consider a single origin server located in each ONOS domain. The main idea is to serve client hosts with the most up-to-date copies of the requested content within a reasonable time (low latency).

More specifically, our application consists of two main components: an `ApplicationManager` and a `DistributedApplicationStore`. The Application Manager component, which is an implementation of the Application Service, is responsible for creating a virtual network of cache servers and providing mesh connectivity between these server hosts. On the other hand, the Distributed Application Store, which is an implementation of the Application Store, performs the task of persisting and synchronizing the information received by the application manager. It is backed by an eventually consistent map with eventual consistency guarantees for storing the service's application state, namely the list of origin servers in the network and their respective set of generated contents:

```
EventuallyConsistentMap < OriginServerID, Set<Content> >
```

Each controller replica that is responsible for a given ONOS domain operates on a local view of the eventually consistent map. This view consists of the local origin server from the same ONOS domain with its generated set of contents, and other potential origin server hosts located in different ONOS domains in the network with their respective set of contents, as seen by the local replica after application state synchronization.

Furthermore, we design a cached map that is local to each controller application instance and that represents the contents cached in the local CDN server within the same ONOS domain. The local cached map is closely linked to the local view of the eventually consistent map, and it reflects the contents stored in the local CDN server. The latter performs the functions of an origin server and at the same time a cache server. Indeed, it contains the contents created locally (the origin server), and potentially other contents that are replicated from other origin servers (the cache server).

More specifically, on a local controller replica, updates to the eventually consistent state map (e.g PUT) might trigger specific actions to feed the local CDN server and consequently update the local cached map. If the update to the content is associated in the map with the local origin server, this means the updated content has already been generated on that origin server. On the other hand, if the update to the content is associated in the map with another origin server from another ONOS domain, our application checks the relevance of that content. In case the content is important to our application, then the update to the content is automatically pulled from the origin server to the local CDN server (cache server) and is cached in the local cached map.

```
CachedMap < ContentName, Set<Content> >
```

5.4.2. *State synchronization and content distribution*

The custom eventually consistent map we use for the synchronization of our CDN application state is based on our own implementation of the EventuallyConsistentMap<K,V> distributed primitive. Indeed, the new implementation we propose for the eventual consistency map abstraction models the Quorum-inspired consistency discussed in 5.3.2.1.

In particular, it takes into account the size of the write Quorum parameter (W) when replicating the updates related to our application's eventually consistent map among the controllers (see Figure 5.4). On each local replica, updates to the local map are queued in time to different EventAccumulators allocated for different controller peers. The latter are selected randomly and their number depends on the write Quorum size W. Whenever an event accumulator is triggered to process the previously accumulated events and propagate them to the associated peer, that peer is removed from the list of Quorum peers. New updates will immediately trigger the creation of a new accumulator associated with a new randomly selected peer that is added to the list of Quorum peers. This accumulator will collect the updates together with the other event accumulators associated with the rest of the Quorum peers. Thus, we guarantee that updates to the eventually consistent map on a local replica are replicated at runtime to exactly W replicas, including the local replica.

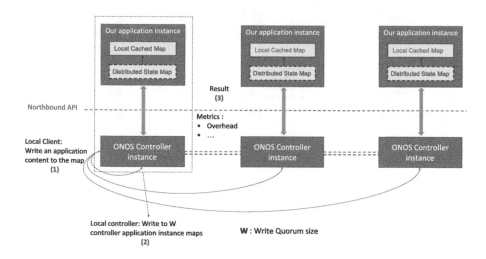

Figure 5.4. *Quorum-inspired Write operations in our CDN-like application. For a color version of this figure, see www.iste.co.uk/bannour/software.zip*

As explained in section 5.4.1, such updates to the eventually consistent map on a local controller replica trigger specific actions that may feed the local CDN server with new contents (content distribution) and thus update the local cached map for our CDN application.

5.4.3. *Content delivery to customers*

During a read operation performed by a client, our controller application instance running on the local controller replica within the same ONOS domain as that client, receives the read request to be fulfilled following Quorum-inspired read consistency protocols (see Figure 5.5).

More specifically, if the read consistency level is higher than ONE (read Quorum size R greater than 1), then the local controller node, which in our case serves as the coordinator node, sends the read request to the remaining randomly selected controller replicas forming the read Quorum. The size R of the read Quorum including the local controller replica is set in advance by the read consistency level.

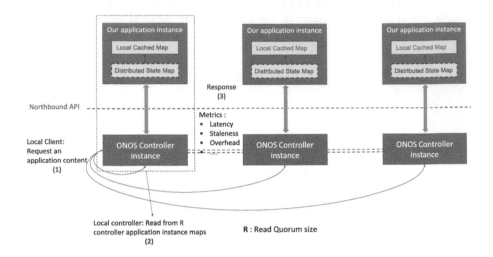

Figure 5.5. *Quorum-inspired read operations in our CDN-like application. For a color version of this figure, see www.iste.co.uk/bannour/software.zip*

We use ONOS's ClusterCommunicationService to assist communications between the local controller node and the rest of the controller cluster nodes in the read Quorum. More specifically, the local controller node sends the read request message with a particular subject to each of the concerned controller nodes using the sendAndReceive method of the cluster communication service. It expects a future reply message from each of the involved controllers that have already subscribed to the same message subject.

That said, to serve the client's read request for a specific content (ContentName), each controller node that has subscribed to the specified message subject receives the read request and uses the application's handler function for processing the incoming message. Accordingly, the application instance on each controller replica of the read Quorum (including the local replica) consults the local cached map. As explained in section 5.4.1, the cached map represents the list of contents (created by different origin servers) being observed in the local view of the eventually consistent map, and then pulled to be cached in the local CDN server. Using that map, each application instance compares the cached versions of the requested content (ContentName) based on their LogicalTimestamp properties, in order to determine the freshest version of the content. Then, it produces a reply containing the selected Content with its four properties, discussed in section 5.4.1, and more importantly the IP address of the local cache server that has just delivered the requested content.

Each content that is created on the origin server, and then eventually propagated to cache servers has four properties: a ContentName, an identifier ID, a real time-based CreationTime, a LogicalTimestamp and a Version.

The local controller replica, playing the role of the coordinator, waits for the read Quorum of replicas to respond. Then, it merges the R responses (including the response produced on the local replica) to figure out the location of the freshest version of the requested content among the concerned CDN servers (equal to R in our scenario). Finally, it sends back the final response to the client and makes sure a host-to-host connectivity intent is added between the client host and the determined cache server host, using the ONOS Intent Framework. Based on this response, the client, which has issued an HTTP request specifying the URL of the requested content, is redirected, using our CDN-like strategy (described above) and a DNS resolution service, to the selected cache server in order to retrieve the specified version of the content.

After each client request, our application collects the continuous consistency and performance metrics related to that request. These metrics are described in detail in section 5.5.1.

5.5. Performance evaluation

5.5.1. *Application-specific performance and consistency metrics*

Here, we show the considered continuous and SLA-style performance and consistency metrics. More specifically, we show how we measure these metrics when implementing our adaptive consistency strategy for the C-CDN application that we designed on top of the distributed ONOS controllers.

– *Performance metrics*

- Network-related metrics: we regard the application inter-controller overhead as a network performance metric. We first capture all inter-controller traffic using TCP port 9876. Then, we filter the captured traffic based on different conditions in order to evaluate the application's inter-controller overhead.

Our goal is to minimize the application overhead due to write and read operations, depending on the given application SLA, the application workload and the network context.

$$AppOverhead = WriteOverhead + ReadOverhead \qquad [5.2]$$

- Client-centric metrics: we also consider the response time to a client request as a performance metric. As defined by our application, the response time consists of the delay to fetch the appropriate version of the requested content from the local cached maps of the application instances running on the R controller replicas of the read Quorum (Latency1), and the delay to retrieve the specified version of the content from the selected cache server host (Latency2). We also note that these latency times do not overlap.

$$ResponseTime = Latency1 + Latency2 \qquad [5.3]$$

– *Consistency metrics*

As explained in section 5.3.1, we consider the application-specific staleness metric from a strictly time-based perspective: it describes the age of the information in terms of wall-clock time. Accordingly, the staleness of the application content C being returned by a read operation at time t is measured as follows:

$$Staleness(C) = QueryTime - CreationTime(C) \qquad [5.4]$$

Moreover, we set the staleness ranges used in the consistency SLA based on the application content refresh rate.

5.5.2. *Experimental setup*

Our experiments are performed on an Ubuntu 18.04 LTS server using ONOS 1.13. We also use Mininet 2.2.1 and an ONOS-provided script (*onos.py*) to start an emulated ONOS network on a single development machine, including a logically centralized ONOS cluster, a modeled control network and a data network. Wireshark is used as a sniffer to capture the inter-controller traffic which uses TCP port 9876.

5.5.2.1. *TCL-Expect scripts*

In this section, we test our proposed adaptive consistency approach explained in sections 5.3 and 5.4, which we will subsequently refer to as ONOS-WAQIC (ONOS-With Adaptive Quorum-Inspired Consistency) for brevity. The proposed approach was implemented for our CDN-like application on ONOS-WAQIC.

To that end, we wrote two Expect Tcl-based scripts (main.exp and onos.exp). In each script, we specified a set of required steps to follow to automate the tasks for our test scenarios on ONOS-WAQIC, as summarized below.

1) First, we ran our startup Expect script (main.exp). With Mininet and *onos.py*, we started up an ONOS cluster and a modeled data network for the specified topology. The selected number N of the ONOS controller replicas that will form the ONOS cluster was passed as an argument to the executed script.

2) Then, we ran the Mininet CLI built-in `pingall` command to discover the network topology. We also launched a spawned process to install and activate the CDN-like application that we developed on ONOS-WAQIC. To force device/switch mastership re-balancing, we connected to one of the running ONOS controller instances, and launched the ONOS CLI `balance-masters` command.

3) First, we parsed the output of the `dump` Mininet command using regular expressions in Tcl in order to build a key-value array mapping the IP addresses of hosts to their Mininet names *(array1)*. Then, in the main Expect script, we launched N spawned processes that connect to the N running ONOS controller instances using the same Expect script (`onos.exp`) that we developed but run with different arguments (controller IP address, content name, maximum number of content versions). In the `onos.exp` script, we analyzed the output of the `masters` ONOS CLI command to construct an array mapping each controller IP with the set of associated switches (MAC IDs) *(array2)*. In addition, using the output of the Mininet CLI `hosts` command, we constructed two additional arrays: the first array associates each host MAC ID with its IP address *(array3)*, and the second array associates each host MAC ID with the switch ID to which it is connected *(array4)*.

4) It is worth noting that our `onos.exp` script starts by running two ONOS CLI commands (`set-read` and `set-write`). We created these commands to set the read and write Quorum sizes R and W to the values specified by the consistency level for a given ONOS controller instance. These values were passed as command arguments.

5) Using *array2*, *array3* and *array4*, each of the N currently spawned processes running the `onos.exp` script for a specific ONOS instance builds another Tcl array *(array5)* that identifies the list of hosts (MAC addresses) associated with each ONOS controller instance (controller IP address) in the network. Based on that array, our script randomly selects, for the specified ONOS controller instance, a list of hosts that will serve as origin cache servers and a list of hosts that will serve as clients in the concerned ONOS controller domain. The number of selected cache and client hosts depends on the application scenario/workload (see section 5.3.2.2). Each ONOS process communicates the MAC and IP addresses of the origin server to the local application instance using our ONOS CLI `set-cache` command. Our script also runs the ONOS CLI `add-host` command which we created to add the cache server hosts to our application's `EventuallyConsistentMap` (discussed in section 5.4.1).

Moreover, information about these cache server hosts is sent (using "puts") to the running `main.exp` script process. The latter identifies the Mininet names of these hosts using *array1* and connects to the Mininet CLI command in order to install a `SimpleHTTPServer` on each of the cache server hosts.

6) At this stage, we ensured that our main process (running *main.exp*) and the N spawned processes (running `onos.exp` with different arguments) were synchronized. Afterwards, each of the N spawned processes connecting to an ONOS controller instance starts adding (then updating with a certain *refresh rate*) the contents to the

origin server host in the involved ONOS domain. We used the add-content ONOS CLI command that we created to add a given content version (second command argument) to the specified origin server host (first command argument) in the application's EventuallyConsistentMap. Further details about content distribution and state synchronization using Quorum-inspired write consistency are provided in section 5.4.2.

On the other hand, in parallel with the updating of contents, our main process, handled by the main.exp script, starts issuing and serving client requests for specific contents. This was achieved using our get-IP-content CLI command, which takes one argument, namely the requested ContentName, and returns the IP address of the cache server containing the freshest/selected version of the requested content. Then, our script retrieves the content from the determined server using "wget". In addition, after each client request, continuous application-specific consistency and performance metrics related to that request are collected with our script using regular expressions in Tcl. More details about the content delivery strategy we follow using Quorum-inspired read consistency are given in section 5.4.3.

5.5.2.2. *OpenAI Gym simulator*

To implement the Machine Learning Module (see section 5.3.2.2.2) for our CDN-like application on ONOS-WAQIC, we build a simulator based on OpenAI Gym, an open-source Python toolkit for developing and comparing reinforcement learning algorithms.

More specifically, we build a new environment to simulate knowledge exchange in an ONOS SDN cluster. We start by building an offline data set using our TCL-Expect scripts, explained in section 5.5.2.1. Our data set stores the information collected by the Monitoring Module about client requests for specific CDN contents. As detailed in section 5.4.3, for a given client request, the returned information contains the current values of the Quorum parameters R and W, the expected returned version of the content (content update step), the actual returned version of the content, the *staleness* of the returned content, the *delay* incurred in searching for the freshest version of the content from R controller replicas (latency1), the *read overhead*, the *write overhead*, and the application scenario determined by the Workload Identifier Module.

The data set is fed to the Automatic Module which hands it over to the Machine Learning Module to learn the read and write Quorum size parameters online. Implemented with Gym, the latter module uses the data set to learn the Kernel Density Estimation (KDE with scipy) for each metric using the data of some clients. These client data are selected with respect to the current configuration of R and W parameters. This configuration was set following an action performed by the agent (see the explanation of the QL algorithm in section 5.3.2.2.2 for more details). Thus, using KDE, our ML Module estimates the expected metrics for each selected

Quorum configuration, and then updates the Q-table with the Q-value of that action at that state, at each step (or episode) of the QL algorithm.

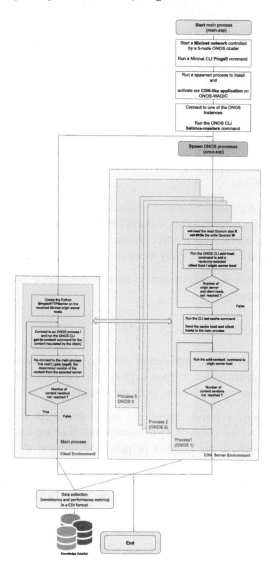

Figure 5.6. *Overview of the main tasks executed by our TCL-Expect scripts. For a color version of this figure, see www.iste.co.uk/bannour/software.zip*

5.5.2.3. *Various learning agent policies*

We implemented three learning agents that adopt different policies. The latter are compared and validated through five scenarios. Each scenario reflects a specific use case (e.g. a latency-sensitive application, a consistency-favoring application). To minimize the application's overall inter-controller overheads, our agents use the estimated overhead as a negative "reward" when performing actions (setting R and W) that change the environment state. The controlled and constrained agents are proposed with the aim of improving the simple greedy agent. Below is a brief description of these agents.

– *A simple ε-greedy agent* (Tran et al. 2019): this agent follows a simple ε-greedy policy with a fixed ϵ value, where ϵ is the exploration rate and (1-ϵ) is the exploitation rate. We test three simple ε-agents: ε-greedy5 ($\epsilon = 0.5$), ε-greedy10 ($\epsilon = 0.10$) and ε-greedy15 ($\epsilon = 0.15$).

– *A controlled dynamic ε-greedy agent*: this agent follows a dynamic ε-greedy strategy where the exploration rate ϵ decays as the algorithm's episode count increases. The purpose is to account for the fact that the agent learns more about the environment in time, and becomes progressively more confident and "greedy" for exploitation.

We use the following decay function for reducing ϵ as a function of episode count. x is the episode number.

$$f(x) = \epsilon * (0.5 + \log_{10}(2 - arctan(\frac{x}{10} - 2))) \qquad [5.5]$$

To attempt to satisfy the application's latency and staleness thresholds, the simple and controlled agents reject, at each exploitation episode, any action violating these constraints and remove its Q-value from the Q-table.

– *A constrained ε-greedy agent*: to make the agent learn how to satisfy the application's SLA, we create a Q-constraint list that we update over the episodes. Its size corresponds to the number of potential actions: the number of R and W combinations such that $R + W \leq N$. The list represents the number of constraint violations by each Quorum configuration. The considered constraints are both the latency and staleness thresholds specified in the SLA. During each exploitation phase, we update the Q-constraint list and use it to generate a new Q-list containing the Quorum configurations that give less constraint violations. These configurations are then exploited. They are compared using their Q-values in the Q-table (based on the estimated overhead reward) to select the best Quorum configuration (action) at that episode.

5.5.3. *Results*

5.5.3.1. *Impact of the read and write Quorum sizes*

In this section, we present an experimental study that is aimed at assessing the impact of using different read and write Quorum sizes (R and W respectively) on the read and write inter-controller overheads of our CDN-like application running on a 5-node ONOS cluster in the network topology.

In the conducted experiments, we consider three application workloads that are representative of three application scenarios (see the Workload Identifier Module in section 5.3.2.2 for more details). For the studied workloads, we show the captured read and write packets within a specified time interval (i.e. 400 ms in our tests) of read and write client operation accesses, for all possible eventually consistent partial Quorum configurations (R,W) (e.g. (R,W) combinations such that $R + W \leq N$ where $N = 5$).

Our results clearly show that, when increasing the read Quorum size R, the number of read packets increases, mainly in a read-dominated workload (see Figure 5.7(a)). In addition, increasing the write Quorum size W results in a drastic increase in the number of write packets, especially in a write-intensive workload. Indeed, their number reaches 400 during the specified time interval for a partial Quorum configuration where W is equal to 5 (see Figure 5.8(b)).

Given the high inter-controller overheads observed in our experimental data for certain Quorum configurations, we propose to tune the R and W parameters and therefore optimize the configuration of the Quorum system to better match the varying application SLA requirements and the dynamic application workloads, as we further discuss in the following sections.

5.5.3.2. *Quorum configuration optimization*

5.5.3.2.1. Dynamic application SLA requirements

To evaluate our ONOS-WAQIC proposal for the CDN-like application we developed, we ran our TCL-Expect scripts (see section 5.5.2.1) with a 5-node ONOS cluster according to different scenarios. In these scenarios, we used different partial Quorum configurations (R, W) and followed various application workloads with respect to different ratios between read and write operations. Then, we used the data collected as an input to our QL simulator (see section 5.5.2.2). In the simulator environment, we set $\alpha = \gamma = 0.5$ and the number of episodes to 1,000. We also consider different test scenarios that reflect different application requirements in terms of performance and consistency, as summarized in Table 5.1.

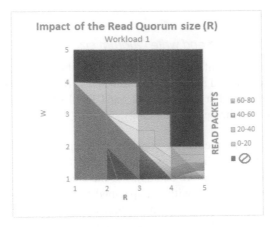

a) Read packets when varying (R,W)

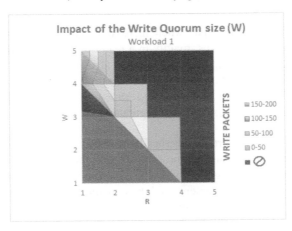

b) Write packets when varying (R,W)

Figure 5.7. *Workload 1: a read-intensive application scenario. For a color version of this figure, see www.iste.co.uk/bannour/software.zip*

In particular, using our data set and knowing the refresh rate of our CDN-like application, we learn the t_staleness ranges. In other words, we learn the relationship between the t_staleness value of a certain content being returned and how old that returned content is by version. As a result, estimating the t_staleness ranges allowed us to set the time-based staleness thresholds in the SLA while having an idea of the associated version-based staleness thresholds.

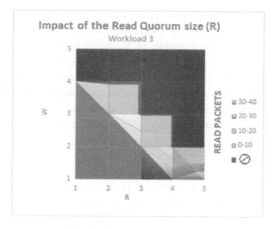

a) Read packets when varying (R,W)

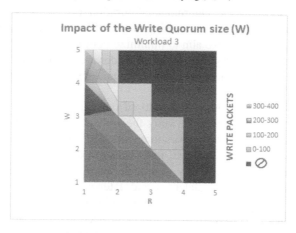

b) Write packets when varying (R,W)

Figure 5.8. *Workload 3: a write-intensive application scenario. For a color version of this figure, see www.iste.co.uk/bannour/software.zip*

Test scenarios	Latency threshold (ms)	t_Staleness threshold (ms)	k_Staleness Version old
No. 1	5	300,000	3
No. 2	25	220,000	2
No. 3	50	120,000	1

Table 5.1. *Application SLA scenarios*

In each test scenario that we run on the simulator, our application expresses the performance and consistency SLAs using the latency threshold (in ms) and the staleness threshold (in ms). For example, in Scenario no. 3 our application, which is consistency-favoring, enforces the following SLA: it expects that a read operation obtains a reply in under 100 ms, and returns a content value no older than 120 s (i.e. no older than 1 version stale). Accordingly, our consistency approach attempts to find the best Quorum combination of R and W so as to minimize the application's read and/or the write inter-controller overheads while ensuring the desired performance-consistency trade-offs.

For a given Quorum configuration, we compute the read overhead ratio by normalizing the generated read overhead (bytes/s) with respect to the Quorum configuration generating the maximum read overhead and zero write overhead (the configuration $(R = 5, W = 1)$) in our case) for each application scenario. We follow the same steps for computing the write overhead ratio based on the generated write overhead with respect to the Quorum configuration $(R = 1, W = 5)$, which corresponds to the standard implementation of ONOS's eventual consistency model. On the other hand, whenever we aim to minimize both the read and write overheads (e.g. in a balanced workload scenario), we consider the mean of the read and write overhead ratios which we will subsequently refer to as the global overhead ratio.

In Figures 5.9–5.11, we show the results of our experimental tests for the three considered application scenarios. To study the impact of changing the application SLA requirements, we set the application workload to Workload 2 (a balanced workload scenario that has a balanced ratio between the read and write operation accesses) in which our consistency approach attempts to minimize the global overhead ratio and satisfy the staleness and latency SLA thresholds set by the application. Moving from one application workload scenario to another (e.g. a read-intensive scenario) is covered in the following section.

Figure 5.9 shows that, in a latency-sensitive application scenario, the constrained and the controlled agent policies are the most appropriate. The number of constraint violations decreases with episode stages (see Figures 5.9(a) and 5.9(b)), and the generated global (read and write) inter-controller overhead (see Figure 5.9(c)) is minimal compared to the simple greedy agent policy, and to the standard ONOS implementation. We also notice that the three agents converge toward Quorum configurations where $R = 1$ (i.e. $(R = 1, W = 2)$, $(R = 1, W = 3)$ and $(R = 1, W = 4)$). This is due to the given strong constraint on latency.

Figure 5.10 shows that, in a balanced application scenario, the constrained and the controlled agent policies offer the best real-time trade-offs between the application's latency and staleness needs (see Figures 5.10(a) and 5.10(b)) while ensuring minimal global overhead ratio (approximately 25%) (see Figure 5.10(c)). In particular, the constrained agent converges toward balanced Quorum configurations

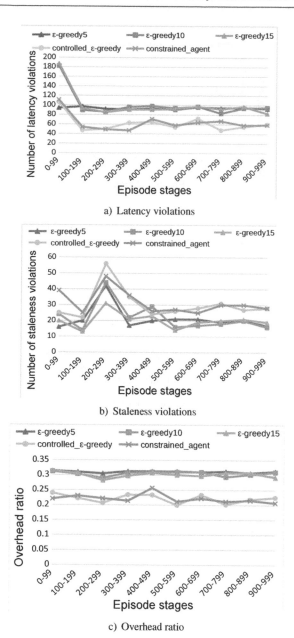

a) Latency violations

b) Staleness violations

c) Overhead ratio

Figure 5.9. *Scenario 1: a latency-sensitive application. For a color version of this figure, see www.iste.co.uk/bannour/software.zip*

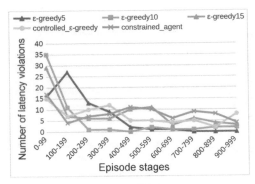

a) Latency violations

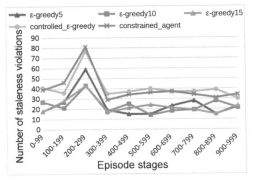

b) Staleness violations

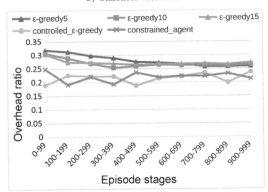

c) Overhead ratio

Figure 5.10. *Scenario 2: a consistency/latency-balancing application. For a color version of this figure, see www.iste.co.uk/bannour/software.zip*

(i.e. $(R = 2, W = 2)$ and $(R = 2, W = 3)$). On the other hand, the simple ϵ-greedy agents provide a small number of latency violations but at the cost of generating more overhead.

As can be seen from Figure 5.11, in a consistency-favoring application scenario, all agents perform well at reducing the staleness violations (see Figure 5.11(b)), especially the simple greedy agents. Moreover, all agents respect the relaxed latency constraint (see Figure 5.11(a)). They all converge toward a common Quorum configuration $(R = 3, W = 2)$. We also note that the constrained and controlled agents ensure a significant gain in overhead, almost 80%.

Other application scenarios were tested, such as Scenario no. 4, in which latency is favored and consistency is completely relaxed ("any"). Our results showed that, in such scenarios, the learning agents converge toward a common Quorum configuration $(R = 1, W = 1)$.

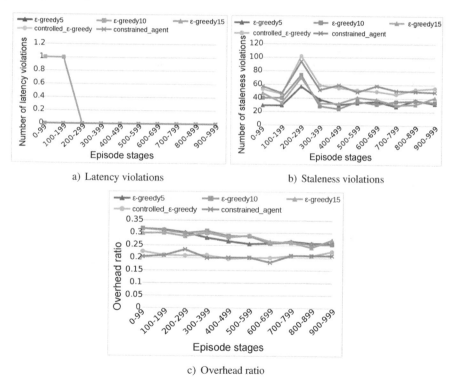

a) Latency violations   b) Staleness violations

c) Overhead ratio

Figure 5.11. *Scenario 3: a consistency-favoring application. For a color version of this figure, see www.iste.co.uk/bannour/software.zip*

Table 5.2 summarizes the final results of the constrained and controlled agents for the considered application scenarios. More specifically, it shows the optimal Quorum configurations (R, W) reached after algorithm convergence for different application SLA requirements.

Application scenario	Latency sensitive	Staleness favoring	Read Quorum size R	Write Quorum size W
No. 1	+++	+	1	3
No. 2	++	++	2	2
No. 3	+	+++	3	2
No. 4	++++	−	1	1

Table 5.2. *Final Q-learning results of the constrained and controlled agents for the considered application scenarios*

5.5.3.2.2. Dynamic application workloads

In this section, we aim to assess the ability of our adaptive Quorum-inspired consistency strategy (ONOS-WAQIC), for the CDN-like application we developed, to adapt to time-varying application workloads. The dynamic changes in such application workload patterns may indeed affect the observed network and application metrics (e.g. inter-controller overhead, staleness and access latency).

Taking this into consideration, our adaptive consistency model attempts to adjust the consistency level at runtime by continuously tuning the Quorum configuration parameters, in order to better match the varying workloads.

In this context, we consider three workloads, as discussed in section 5.3.2.2 (see the Workload Identifier Module). In the three studied workloads, our model aims to satisfy the latency and staleness SLA requirements. Additionally, in the read-dominated workload (Workload 1), our model attempts to minimize the read overhead. Conversely, in a write-intensive workload (Workload 3), it focuses on reducing the write overhead. Finally, in a balanced workload, our approach aims to minimize both the read and write overheads (the global overhead).

To experiment with these workloads, we fix the application scenario to 2 (see section 5.5.3.2.1) which represents an application scenario with balanced consistency (staleness)/latency SLA requirements. Then, we conduct some tests on our QL simulator. During these tests, we apply different variations in the application workload. More specifically, the first time period (the first 400 episodes) is characterized by a balanced workload (Workload 2). At episode 400, we run a read-dominated workload (Workload 1). Finally, starting from episode 700, we consider a write-intensive workload (Workload 3).

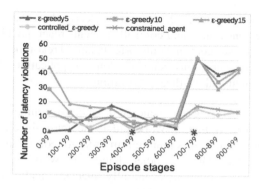

a) Latency violations

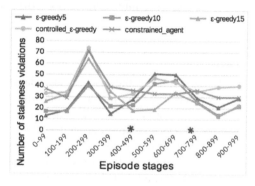

b) Staleness violations

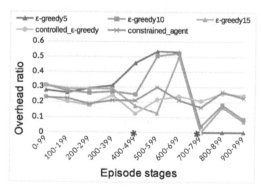

c) Overhead ratio

Figure 5.12. *Dynamic changes in the workload (Workload 2-Workload 1-Workload 3) in a consistency/latency-balancing application scenario (Scenario no. 2). For a color version of this figure, see www.iste.co.uk/bannour/software.zip*

As can be seen from Figure 5.12, our results clearly show that, unlike the simple ϵ-greedy agents, the constrained and controlled agents react quickly to the dynamic workload variations. These agents not only offer balanced real-time trade-offs between the performance (latency) and consistency (staleness) application SLA requirements, but they also provide minimal overhead at runtime.

Moreover, when analyzing the generated Quorum configurations during the conducted tests, we observe that, in Workload 1, the constrained agent converges to Quorum configurations where R is minimal, in order to reduce the application's read inter-controller overhead. On the other hand, in Workload 3, the Quorum configurations where W is small are eventually selected. Finally, in Workload 2, the constrained agent converges to balanced Quorum configurations in which $R = W = 2$ to reduce the application's read and write inter-controller overheads.

5.6. Conclusion

In this chapter, we further studied the use of an adaptive and continuous consistency model for the distributed SDN controllers, following the notion of partial Quorum consistency at scale (Bannour et al. 2019, 2020). Our consistency adaptation strategy was implemented for a CDN-like application developed on top of ONOS. It mainly consists of turning ONOS's optimistic replication technique into a more scalable and intelligent Quorum-inspired replication strategy using various online QL RL approaches. Our experiments showed that the constrained ϵ-greedy approach we tested in a 5-node ONOS cluster proved efficient in helping our C-CDN-like application find the appropriate read and write Quorum replication parameters at runtime. In fact, the adjustable and time-varying partial Quorum configurations determined by our strategy at runtime have, under changing network and application workload conditions, achieved balanced trade-offs between the application's continuous performance (latency) and consistency (staleness) requirements. Moreover, these real-time trade-offs ensured a substantial reduction in the application's inter-controller read and write overhead (especially in a large-scale ONOS network) while satisfying the application-defined thresholds specified in the given application SLA.

Our proposed adaptive and Quorum-inspired consistency model could be further enhanced by leveraging the compulsory anti-entropy reconciliation mechanisms proposed in Bannour et al. (2018a) (*expanding partial Quorums*) (see Chapter 4). Such mechanisms are indeed useful in particular cases (e.g. failure scenarios, controller crashes) where the system consistency observed by the applications is at high risk, and cannot be fixed by only adjusting the Quorum parameters.

Our proposed CDN-like application could evidently be leveraged by several ICN use-cases. For example, it can meet the needs of video-on-demand services and, more particularly, the requirements of live and pre-recorded streaming where video files are versioned based on current time (Kulinski and Burke 2012). Another interesting use-case could be vehicular networks. Our proposal can indeed provide solutions to frequent changes in network topology state and content (e.g. degree of pollution and traffic conditions) (Kalogeiton et al. 2017).

Finally, it is worth noting that our self-adaptive and automated consistency mechanisms for the distributed SDN controllers could be applied to many other concrete distributed network applications. The latter should have consistency adaptability requirements at scale, such as the need for a dynamic adaptation of the replication style given the varying patterns of application behavior and network context. These applications could include Cloud data storage services, Website visitors (e.g. discussion forums), e-commerce and media-service providers (e.g. Amazon and Netflix), security applications, as well as smart-home services.

Conclusions and Perspectives

C.1. Summary of contributions

Software-defined networking (SDN) has increasingly gained traction over the past few years in both academia and research. The SDN paradigm builds its promises on the separation of concerns between the network control logic and the forwarding devices, as well as the logical centralization of the network intelligence in software components. Becuase of these key attributes, SDN is believed to work with network virtualization to fundamentally change the networking landscape toward more flexible, agile, adaptable and highly automated next-generation networks.

Despite all the hype, SDN raises many concerns and questions regarding its implementation and deployment. For instance, current SDN deployments based on physically centralized control architectures present several issues of scalability and reliability. As a result, distributed SDN control architectures were proposed as a suitable solution for overcoming these problems. However, there are still ongoing community debates about the best and most optimal approach to decentralizing the network control plane in order to harness the full potential of SDN.

In the early stages of our work, we conducted a comprehensive literature survey of the wide variety of existing SDN controller platforms. In addition to reviewing the SDN concept and comparing the SDN architecture to the traditional network architecture, we proposed a taxonomy of state-of-the-art SDN controller platforms by categorizing them in two ways: based on a physical classification or a logical classification. Our thorough study of these SDN platform proposals allowed us to shed more light on their advantages and drawbacks and to develop a critical awareness of the challenges facing distributed control in SDNs.

Scalability, reliability, consistency and interoperability of the SDN control plane are among the key competing challenges encountered in designing an efficient and robust high-performance distributed SDN controller platform. Although considered the main limitations of fully centralized SDN control designs, scalability and reliability are also major concerns that are expressed in the context of distributed SDN architectures. They are indeed heavily impacted by the structure of the distributed control plane (e.g. flat, hierarchical or hybrid organization) as well as the number and placement of the multiple controllers within the SDN network. Achieving such performance and availability requirements usually comes at the cost of guaranteeing a consistent centralized network view that is required for the correct behavior of SDN applications. Consistency considerations should therefore be explored among the trade-offs involved in the design process of a decentralized SDN controller platform.

Given the wide range of promising SDN controller platforms, each with their own set of major challenges, we argue that developing a brand new one may not be the best solution. However, it is essential to leverage the existing platforms by aggregating, merging and improving their proposed ideas in order to get as close as possible to a common standard that could emerge in the coming years. This distributed SDN controller platform should meet the emerging challenges associated with large-scale deployments and, most importantly, with next-generation networks (e.g. IoT (Ojo et al. 2016) and fog computing (Liang et al. 2017)).

With these considerations in mind, in the further stages of our work we intended to address some of the previously discussed challenges associated with the complex problem of designing a distributed SDN control plane. To that end, we proposed to split that problem into two manageable challenges that are correlated: the controller placement problem (1) and the knowledge sharing problem (2). The first problem investigates the required number of controllers, along with their appropriate locations with respect to the desired performance and reliability objectives and depending on the existing constraints. The second problem is related to the type and amount of information to be shared among the controller instances given a desired level of consistency.

First, we addressed the SDN controller placement optimization problem in the context of large-scale IoT-like networks. To that end, we put forward four scalable strategies covering different aspects of the multi-objective controller placement optimization problem, with respect to multiple reliability and performance metrics that are considered according to different uses and contexts. To assess these strategies, two heuristic-based approaches were proposed with the objective of finding high-quality approximate solutions to the controller placement problem in a reasonable computation time: a clustering approach (PAM-B) based on a dissimilarity score and a modified genetic approach (NSGA-II). Our results demonstrated the potential of clustering techniques in delivering appropriate controller placement configurations that achieve balanced trade-offs among the competing performance and reliability criteria at scale.

We then investigated the knowledge dissemination problem between the distributed SDN controllers by proposing an adaptive multi-level consistency model following the notion of continuous consistency for the distributed SDN controllers. This model provides many advantages for SDN applications when compared to the strong consistency and eventual consistency extremes, especially in large-scale deployments. It delivers the scalability, performance and availability benefits of an eventual consistency model but has the additional advantage of controlling the observed state inconsistencies in an application-specific manner. More specifically, we proposed two different scalable consistency approaches for the open-source ONOS controllers and compared them with ONOS's static strategies to eventual state consistency.

The first consistency approach was implemented for a source routing application on top of ONOS. It involved turning ONOS's eventual consistency model into an adaptive consistency model using the *anti-entropy reconciliation period* as a *control knob* for an adaptive fine-grained tuning of consistency levels. In addition to ensuring the application's continuous consistency requirements (i.e. *Numerical Error* bounds) as specified in the given application SLA, our results showed a substantial reduction in the anti-entropy reconciliation overhead compared to ONOS's static consistency scheme at scale.

The second approach extended the adaptive consistency strategy to the optimistic replication technique used in ONOS's eventual consistency model. It was implemented for a CDN-like application we developed on top of the ONOS controllers. It mainly involved turning ONOS's optimistic replication technique into a more scalable and intelligent Quorum-inspired replication strategy using various Q-learning RL approaches. In particular, it uses the *read and write partial Quorum parameters* as adjustable *control knobs* for a fine-grained consistency tuning, rather than relying on anti-entropy reconciliation mechanisms. Our experiments showed that the proposed constrained ϵ-greedy approach proved efficient in finding the appropriate read and write Quorum replication parameters at runtime that, under changing network and workload conditions, achieve balanced trade-offs between the application's continuous performance (*latency*) and consistency (*staleness*) requirements. These real-time trade-offs ensured a significant reduction in the application overhead while satisfying the application requirements specified in the SLA.

C.2. Perspectives and future work

Based on the promising results of the work presented in this book, the study can be further extended with a variety of research perspectives.

– First, the controller placement strategies proposed in Chapter 3 could be further enhanced by including a *dynamic controller placement policy*. The latter should take into account the dynamic nature of the network such as the network load or a dynamic network topology.

– Our adaptive consistency model proposed for the distributed ONOS controllers reduces the application inter-controller overhead and tunes the consistency level at runtime, in order to achieve – under changing application workload conditions – balanced real-time trade-offs between the application's continuous performance and consistency requirements (as specified in the given SLA). Other important factors to be considered as part of our future work include the *changing network conditions* in the case of in-bound SDN control.

– The adaptive consistency strategies proposed in this work dynamically adjust at runtime the same consistency level for all the SDN controller instances in the cluster in order to meet certain network and application requirements. Another potential approach is to assign different consistency levels to the different controllers (granular *per-controller consistency*) depending on application requirements. Accordingly, in the case of our Quorum-inspired consistency approach, adjusting the consistency level at runtime would mean assigning different Quorum parameter configurations (R, W) to the considered SDN controller replicas in the cluster.

– The Quorum-inspired consistency strategy presented in this book uses a constant *replication factor* equal to the total number of controllers in the cluster. It would also be interesting to use the replication degree as a tunable configuration parameter (or a control knob) to disseminate the knowledge in specific geographical areas according to various application scopes and needs.

– Although the main focus of this work was on dynamically adjusting the consistency level of SDN application states (which use controller states), the work can be further extended to the controller states (internal controller applications). Indeed, the long-term goal of this work is to design *adaptively consistent controllers* that adjust the consistency levels for both control and application planes under changing network conditions.

– The adaptive and continuous Quorum-inspired consistency approach proposed in this book was implemented for certain types of applications (e.g. a CDN-like application) using a 5-node ONOS controller cluster in a emulated network environment. This helped us to assess the feasibility of our solution in distributed SDN control and develop a critical awareness of the challenges faced. The next step of this work is to develop a more effective proof-of-concept (PoC) for distributed SDN controllers in a production environment. This can be achieved by setting up an SDN test bed using more than five controller instances in the SDN cluster, and by experimenting with a variety of industrial use cases in order to test the performance and the degree of functionality of our approach in large-scale, real-world SDN deployments.

– In our work, the controller placement problem and the knowledge sharing problem between the distributed SDN controllers are regarded as correlated problems but have been addressed separately. It would be interesting to consider the knowledge dissemination challenge (state consistency metrics) when investigating the optimal placement of controllers. For example, minimizing the inter-controller latencies in the controller placement process reduces the cost of inter-controller communications and enhances network consistency and performance.

– Finally, in this book we placed a special focus on tackling the control consistency issues in SDN, and we proposed practical solutions that we applied to current SDN controller platforms. In fact, we believe that it is crucial for a distributed SDN architecture to support fault tolerance and consistency checks in order to ensure an efficient and secure SDN control plane (Abdou et al. 2018). As part of our future work, we propose to further address the data/state consistency challenges from a security perspective. In particular, we highlight the need to secure the communications between the SDN controllers against the potential threats facing the SDN control plane. A straightforward example of these threats is a malicious controller replica that sends erroneous data to compromise the system by harming a particular service or network, or by favoring its actions to the detriment of the rest of the controller replicas. A potential solution is to use a blockchain to provide a guarantee of non-repudiation and non-alteration (integrity) of data. This blockchain can also be used to store smart contracts. These contracts are programs that control the permission of data exchanges between the parties under certain conditions. In particular, smart contracts can enable a fine-grained access control of the knowledge to be shared between the controller replicas by establishing elaborate rules (e.g. by allowing access to a particular knowledge only for specific controller replicas).

References

Abadi, D.J. (2012). Consistency tradeoffs in modern distributed database system design: CAP is only part of the story. *Computer*, 45, 37–42.

Abdou, A., van Oorschot, P.C., Wan, T. (2018). Comparative analysis of control plane security of SDN and conventional networks. *IEEE Communications Surveys & Tutorials*, 20(4), 3542–3559.

Abdullahi, A., Manickam, S., Karuppayah, S. (2021). A review of scalability issues in software-defined exchange point (SDX) approaches: State-of-the-art. *IEEE Access*, 9, 74499–74509.

Abuarqoub, A. (2020). A review of the control plane scalability approaches in software defined networking. *Future Internet*, 12(3), 49.

Ahmad, S. and Mir, A.H. (2021). Scalability, consistency, reliability and security in SDN controllers: A survey of diverse SDN controllers. *Journal of Network and Systems Management*, 29(1), 1–59.

Ahmadi, V., Jalili, A., Khorramizadeh, S.M., Keshtgari, M. (2015). A hybrid NSGA-II for solving multiobjective controller placement in SDN. *Proceedings of the 2nd International Conference on Knowledge-Based Engineering and Innovation (KBEI)*. doi: 10.1109/KBEI.2015.7436122.

Akka Framework (n.d.). Available at: http://akka.io/ [Accessed 15 February 2020].

Alam, I., Sharif, K., Li, F., Latif, Z., Karim, M.M., Biswas, S., Nour, B., Wang, Y. (2020). A survey of network virtualization techniques for internet of things using SDN and NFV. *ACM Computing Surveys (CSUR)*, 53(2), 1–40.

Almadani, B., Beg, A., Mahmoud, A. (2021). DSF: A distributed SDN control plane framework for the East/West interface. *IEEE Access*, 9, 26735–26754.

Amin, R., Reisslein, M., Shah, N. (2018). Hybrid SDN networks: A survey of existing approaches. *IEEE Communications Surveys & Tutorials*, 20(4), 3259–3306.

Amiri, E., Alizadeh, E., Raeisi, K. (2019). An efficient hierarchical distributed SDN controller model. *Proceedings of the 2019 5th Conference on Knowledge-Based Engineering and Innovation (KBEI)*. IEEE.

AMQP (n.d.). Available at: http://www.amqp.org/ [Accessed 5 January 2021].

Anerousis, N., Chemouil, P., Lazar, A.A., Mihai, N., Weinstein, S.B. (2021). The origin and evolution of open programmable networks and SDN. *IEEE Communications Surveys & Tutorials*, 23(3), 1956–1971.

Ansible (n.d.). Available at: https://www.ansible.com/ [Accessed 2 January 2021].

Arashloo, M.T., Koral, Y., Greenberg, M., Rexford, J., Walker, D. (2016). SNAP: Stateful network-wide abstractions for packet processing. *Proceedings of the 2016 ACM SIGCOMM Conference, SIGCOMM '16*. Florianopolis.

Aslan, M. and Matrawy, A. (2016). Adaptive consistency for distributed SDN controllers. *Proceedings of the 2016 17th International Telecommunications Network Strategy and Planning Symposium (Networks)*. doi: 10.1109/NETWKS.2016.7751168.

Aslan, M. and Matrawy, A. (2018). A clustering-based consistency adaptation strategy for distributed SDN controllers. *Proceedings of the 2018 4th IEEE Conference on Network Softwarization and Workshops (NetSoft)*, Montreal.

Azodolmolky, S., Wieder, P., Yahyapour, R. (2013). Performance evaluation of a scalable software-defined networking deployment. *Proceedings of the 2013 2nd European Workshop on Software Defined Networks*. doi: 10.1109/EWSDN.2013.18.

Badotra, S. and Panda, S.N. (2020). Evaluation and comparison of OpenDayLight and open networking operating system in software-defined networking. *Cluster Computing*, 23(2), 1281–1291.

Bailis, P., Venkataraman, S., Franklin, M.J., Hellerstein, J.M., Stoica, I. (2012). Probabilistically bounded staleness for practical partial quorums. *Proceedings of the VLDB Endowment*, 5(8), 776–787.

Bailis, P., Venkataraman, S., Franklin, M.J., Hellerstein, J.M., Stoica, I. (2014). Quantifying eventual consistency with PBS. *Communications of the ACM*, 57(8), 93–102.

Bannour, F., Souihi, S., Mellouk, A. (2017). Scalability and reliability aware SDN controller placement strategies. *Proceedings of the 2017 13th International Conference on Network and Service Management (CNSM)*, Tokyo.

Bannour, F., Souihi, S., Mellouk, A. (2018a). Adaptive state consistency for distributed ONOS controllers. *Proceedings of the 2018 IEEE Global Communications Conference (Globecom)*, Abu Dhabi.

Bannour, F., Souihi, S., Mellouk, A. (2018b). Distributed SDN control: Survey, taxonomy, and challenges. *IEEE Communications Surveys & Tutorials*, 20(1), 333–354.

Bannour, F., Souihi, S., Mellouk, A. (2019). Adaptive quorum-inspired SLA-aware consistency for distributed SDN controllers. *Proceedings of the 2019 15th International Conference on Network and Service Management (CNSM)*. IEEE, 1–7.

Bannour, F., Souihi, S., Mellouk, A. (2020). Adaptive distributed SDN controllers: Application to content-centric delivery networks. *Future Generation Computer Systems*, 113, 78–93.

Bari, M.F., Boutaba, R., Esteves, R.P., Granville, L.Z., Podlesny, M., Rabbani, M.G., Zhang, Q., Zhani, M.F. (2013). Data center network virtualization: A survey. *IEEE Communications Surveys & Tutorials*, 15(2), 909–928.

Benamrane, F., Ben Mamoun, M., Redouane, B. (2015). Performances of openflow-based software-defined networks: An overview. *Journal of Networks*, 10(6), 329–337.

Benson, T., Akella, A., Maltz, D.A. (2010). Network traffic characteristics of data centers in the wild. *Proceedings of the 10th ACM SIGCOMM Conference on Internet Measurement, IMC '10*. ACM, New York.

Berde, P., Gerola, M., Hart, J., Higuchi, Y., Kobayashi, M., Koide, T., Lantz, B., O'Connor, B., Radoslavov, P., Snow, W. et al. (2014). ONOS: Towards an open, distributed SDN OS. *Proceedings of the 3rd Workshop on Hot Topics in Software Defined Networking, HotSDN '14*. ACM, New York.

Bessani, A., Sousa, J., Alchieri, E.E.P. (2014). State machine replication for the masses with BFT-SMART. *Proceedings of the 2014 44th Annual IEEE/IFIP International Conference on Dependable Systems and Networks, DSN '14*.

Bianchi, G., Bonola, M., Capone, A., Cascone, C. (2014). OpenState: Programming platform-independent stateful openflow applications inside the switch. *SIGCOMM Computer Communication Review*, 44(2), 44–51.

Bianchi, G., Bonola, M., Pontarelli, S., Sanvito, D., Capone, A., Cascone, C. (2016). Open Packet Processor: A programmable architecture for wire speed platform-independent stateful in-network processing. *CoRR*, abs/1605.01977.

Bianco, A., Giaccone, P., Domenico, S.D., Zhang, T. (2016). The role of inter-controller traffic for placement of distributed SDN controllers. *CoRR*, abs/1605.09268.

Bifulco, R. and Rétvári, G. (2018). A survey on the programmable data plane: Abstractions, architectures, and open problems. *Proceedings of the 2018 IEEE 19th International Conference on High Performance Switching and Routing (HPSR)*, Bucharest.

Bifulco, R., Boite, J., Bouet, M., Schneider, F. (2016). Improving SDN with InSPired switches. *Proceedings of the Symposium on SDN Research, SOSR '16*. Santa Clara. doi: 10.1145/2890955.2890962.

Blenk, A., Basta, A., Reisslein, M., Kellerer, W. (2016). Survey on network virtualization hypervisors for software defined networking. *IEEE Communications Surveys & Tutorials*, 18(1), 655–685.

Bondkovskii, A., Keeney, J., van der Meer, S., Weber, S. (2016). Qualitative comparison of open-source SDN controllers. *Proceedings of the NOMS 2016–2016 IEEE/IFIP Network Operations and Management Symposium*, Istanbul.

Bosshart, P., Daly, D., Gibb, G., Izzard, M., McKeown, N., Rexford, J., Schlesinger, C., Talayco, D., Vahdat, A., Varghese, G. et al. (2014). P4: Programming Protocol-independent Packet Processors. *SIGCOMM Computer Communication Review*, 44(3), 87–95.

Botelho, F., Bessani, A., Ramos, F.M.V., Ferreira, P. (2014). On the design of practical fault-tolerant SDN controllers. *Proceedings of the 2014 3rd European Workshop on Software Defined Networks*, Washington, DC.

Botelho, F., Ribeiro, T.A., Ferreira, P., Ramos, F.M.V., Bessani, A. (2016). Design and implementation of a consistent data store for a distributed SDN control plane. *Proceedings of the 2016 12th European Dependable Computing Conference (EDCC)*, Gothenburg.

Bouajjani, A., Enea, C., Hamza, J. (2014). Verifying eventual consistency of optimistic replication systems. *Proceedings of the ACM SIGPLAN Symposium on Principles of Programming Languages*, California.

Caesar, M., Caldwell, D., Feamster, N., Rexford, J., Shaikh, A., van der Merwe, J. (2005). Design and implementation of a routing control platform. *Proceedings of the 2nd Conference on Symposium on Networked Systems Design & Implementation – Volume 2, NSDI '05*. USENIX Association, Berkeley.

Canini, M., Cicco, D.D., Kuznetsov, P., Levin, D., Schmid, S., Vissicchio, S. (2014). STN: A robust and distributed SDN control plane. *Proceedings of the Open Networking Summit (ONS) Research Track*, California.

Canini, M., Kuznetsov, P., Levin, D., Schmid, S. (2015). A distributed and robust SDN control plane for transactional network updates. *Proceedings of the 2015 IEEE Conference on Computer Communications (INFOCOM)*. doi: 10.1109/INFOCOM.2015.7218382.

Chandra, T.D., Griesemer, R., Redstone, J. (2007). Paxos made live: An engineering perspective. *Proceedings of the 26th Annual ACM Symposium on Principles of Distributed Computing, PODC '07*. ACM, New York.

Chandrasekaran, B. and Benson, T. (2014). Tolerating SDN application failures with LegoSDN. *Proceedings of the 13th ACM Workshop on Hot Topics in Networks HotNets-XIII*. ACM, Los Angeles.

Chandrasekaran, B., Tschaen, B., Benson, T. (2016). Isolating and tolerating SDN application failures with LegoSDN. *Proceedings of the Symposium on SDN Research, SOSR '16*. Santa Clara.

Chihoub, H.-E., Ibrahim, S., Antoniu, G., Pérez, M.S. (2012). Harmony: Towards automated self-adaptive consistency in Cloud storage. *Proceedings of the 2012 IEEE International Conference on Cluster Computing*. doi: 10.1109/CLUSTER.2012.56.

Chihoub, H.-E., Pérez, M., Antoniu, G., Bougé, L. (2013). Chameleon: Customized application-specific consistency by means of behavior modeling. Research report, hal-00875947.

Chung, J., Cox, J., Ibarra, J., Bezerra, J., Morgan, H., Clark, R., Owen, H. (2015). AtlanticWave-SDX: An international SDX to support science data applications. *Proceedings of the Software Defined Networking (SDN) for Scientific Networking Workshop*. Austin.

Chung, J., Owen, H., Clark, R. (2016). SDX architectures: A qualitative analysis. *Proceedings of the SoutheastCon 2016*. doi: 10.1109/SECON.2016.7506749.

Clark, D.D., Partridge, C., Ramming, J.C., Wroclawski, J.T. (2003). A knowledge plane for the Internet. *Proceedings of the 2003 Conference on Applications, Technologies, Architectures, and Protocols for Computer Communications, SIGCOMM '03*. ACM, New York.

Costa, L.C., Vieira, A.B., de Britto e Silva, E., Macedo, D.F., Vieira, L.F., Vieira, M.A., da Rocha Miranda, M., Batista, G.F., Polizer, A.H., Gonçalves, A.V.G.S. et al. (2021). OpenFlow data planes performance evaluation. *Performance Evaluation*, 147, 102194.

Couceiro, M., Chandrasekara, G., Bravo, M., Hiltunen, M., Romano, P., Rodrigues, L. (2015). Q-OPT: Self-tuning Quorum system for strongly consistent software defined storage. *Proceedings of the 16th Annual Middleware Conference, Middleware '15*. Vancouver.

Curtis, A.R., Mogul, J.C., Tourrilhes, J., Yalagandula, P., Sharma, P., Banerjee, S. (2011). Devoflow: Scaling flow management for high-performance networks. *Proceedings of the ACM SIGCOMM 2011 Conference, SIGCOMM '11*. ACM, New York.

Das, T. and Gurusamy, M. (2020). Controller placement for resilient network state synchronization in multi-controller SDN. *IEEE Communications Letters*, 24(6), 1299–1303.

Das, T., Sridharan, V., Gurusamy, M. (2019). A survey on controller placement in SDN. *IEEE Communications Surveys & Tutorials*. doi: 10.1109/COMST.2019.2935453.

Das, R.K., Pohrmen, F.H., Maji, A.K., Saha, G. (2020). FT-SDN: A fault-tolerant distributed architecture for software defined network. *Wireless Personal Communications*, 114(2), 1045–1066.

Dhawan, M., Poddar, R., Mahajan, K., Mann, V. (2015). SPHINX: Detecting security attacks in software-defined networks [Online]. Available at: https://rishabhpoddar.com/publications/Sphinx.pdf.

Doria, A., Salim, J.H., Haas, R., Khosravi, H., Wang, W., Dong, L., Gopal, R., Halpern, J. (2010). Forwarding and Control Element Separation (ForCES) protocol specification. Internet Engineering Task Force, March.

ENDEAVOUR Project (n.d.). Available at: https://www.h2020-endeavour.eu/ [Accessed 2 January 2021].

Erickson, D. (2013). The beacon openflow controller. *Proceedings of the 2nd ACM SIGCOMM Workshop on Hot Topics in Software Defined Networking, HotSDN '13*. ACM, New York.

Feamster, N., Rexford, J., Zegura, E. (2014). The road to SDN: An intellectual history of programmable networks. *SIGCOMM Computer Communication Review*, 44(2), 87–98.

Floodlight Project (n.d.). Available at: https://floodlight.atlassian.net/wiki/spaces/ floodlightcontroller/overview [Accessed 7 March 2021].

Fonseca, P. and Mota, E. (2017). A survey on fault management in software-defined networks. *IEEE Communications Surveys & Tutorials*, PP(99), 1–1.

Fonseca, P., Bennesby, R., Mota, E., Passito, A. (2013). Resilience of SDNs based on active and passive replication mechanisms. *Proceedings of the 2013 IEEE Global Communications Conference (GLOBECOM)*. doi: 10.1109/GLOCOM. 2013. 6831399.

Foster, N., Harrison, R., Freedman, M.J., Monsanto, C., Rexford, J., Story, A., Walker, D. (2011). Frenetic: A network programming language. *Proceedings of the 16th ACM SIGPLAN International Conference on Functional Programming, ICFP '11*. ACM, New York.

Ganatra, N.S.A. (2012). Comparative study of several clustering algorithms. *International Journal of Advanced Computer Research*, 37–42.

Guang Y., Jun Bi, M.I.Y.L., Guo, L. (2014). On the capacitated controller placement problem in software defined networks. *IEEE Communications Letters*, 18(8), 1339–1342.

Gude, N., Koponen, T., Pettit, J., Pfaff, B., Casado, M., McKeown, N., Shenker, S. (2008). NOX: Towards an operating system for networks. *SIGCOMM Computer Communication Review*, 38(3), 105–110.

Gupta, A., Shahbaz, M., Vanbever, L., Kim, H., Clark, R., Feamster, N., Rexford, J., Shenker, S. (2014). SDX: A software defined internet exchange. *Proceedings of the ACM SIGCOMM*.

Gupta, A., MacDavid, R., Birkner, R., Canini, M., Feamster, N., Rexford, J., Vanbever, L. (2016). An industrial-scale software defined internet exchange point. *Proceedings of the 13th USENIX Symposium on Networked Systems Design and Implementation (NSDI 16)*. USENIX Association, Santa Clara.

Hassas Yeganeh, S. and Ganjali, Y. (2012). Kandoo: A framework for efficient and scalable offloading of control applications. *Proceedings of the 1st Workshop on Hot Topics in Software Defined Networks, HotSDN '12*. ACM, New York.

Heller, B., Sherwood, R., McKeown, N. (2012). The controller placement problem. *Proceedings of the 1st Workshop on Hot Topics in Software Defined Networks, HotSDN '12*. ACM, New York.

Hock, D., Hartmann, M., Gebert, S., Jarschel, M., Zinner, T., Tran-Gia, P. (2013). Pareto-optimal resilient controller placement in SDN-based core networks. *Proceedings of the 25th International Teletraffic Congress (ITC)*. Shanghai.

Hohlfeld, O., Kempf, J., Reisslein, M., Schmid, S., Shah, N. (2018). Guest editorial scalability issues and solutions for software defined networks. *IEEE Journal on Selected Areas in Communications*, 36(12), 2595–2602.

Hong, C.-Y., Kandula, S., Mahajan, R., Zhang, M., Gill, V., Nanduri, M., Wattenhofer, R. (2013). Achieving high utilization with software-driven WAN. *Proceedings of the ACM SIGCOMM 2013 Conference on SIGCOMM, SIGCOMM '13*. ACM, New York.

Hu, Y., Wang, W., Gong, X., Que, X., Cheng, S. (2013). Reliability-aware controller placement for software-defined networks. *Proceedings of the 2013 IFIP/IEEE International Symposium on Integrated Network Management (IM 2013)*. IEEE.

Hunt, P., Konar, M., Junqueira, F.P., Reed, B. (2010). Zookeeper: Wait-free coordination for internet-scale systems. *Proceedings of the 2010 USENIX Conference on USENIX Annual Technical Conference, USENIXATC '10*.

IBM ILOG CPLEX Optimizer (n.d.). Available at: http://www-03.ibm.com/software/products/en/ibmilogcpleoptistud [Accessed 5 June 2017].

Internet2 (n.d.). Advanced layer 2 system. Available at: https://www.internet2.edu/products-services/advanced-networking/layer-2-services/ [Accessed 9 October 2017].

Jain, S., Kumar, A., Mandal, S., Ong, J., Poutievski, L., Singh, A., Venkata, S., Wanderer, J., Zhou, J., Zhu, M. et al. (2013). B4: Experience with a globally-deployed software defined WAN. *Proceedings of the ACM SIGCOMM 2013 Conference on SIGCOMM, SIGCOMM '13*. ACM, New York.

Jalili, A., Keshtgari, M., Akbari, R. (2018). Optimal controller placement in large scale software defined networks based on modified NSGA-II. *Applied Intelligence*, 48(9).

Jammal, M., Singh, T., Shami, A., Asal, R., Li, Y. (2014). Software defined networking: State of the art and research challenges. *Computer Networks*, 72, 74–98.

Jeyakumar, V., Alizadeh, M., Kim, C., Mazières, D. (2013). Tiny packet programs for low-latency network control and monitoring. *Proceedings of the 12th ACM Workshop on Hot Topics in Networks, HotNets-XII*. College Park. doi: 10.1145/2535771.2535780.

Kalogeiton, E., Zhao, Z., Braun, T. (2017). Is SDN the solution for NDN-VANETs? *Proceedings of the 2017 16th Annual Mediterranean Ad Hoc Networking Workshop (Med-Hoc-Net)*, Budva.

Karakus, M. and Durresi, A. (2017). A survey: Control plane scalability issues and approaches in software-defined networking (SDN). *Computer Networks*, 112, 279–293.

Katta, N., Zhang, H., Freedman, M., Rexford, J. (2015). Ravana: Controller fault-tolerance in software-defined networking. *Proceedings of the 1st ACM SIGCOMM Symposium on Software Defined Networking Research, SOSR '15*. Santa Clara.

Khorsandroo, S., Sanchez, A.G., Tosun, A.S., Rodríguez, J.M.A., Doriguzzi-Corin, R. (2021). Hybrid SDN evolution: A comprehensive survey of the state-of-the-art. *Computer Networks*, 192, 107981.

Killi, B.P.R. and Rao, S.V. (2019). Controller placement in software defined networks: A comprehensive survey. *Computer Networks*, 163, 106883.

Kim, H. and Feamster, N. (2013). Improving network management with software defined networking. *IEEE Communications Magazine*, 51(2), 114–119.

Klophaus, R. (2010). Riak core: Building distributed applications without shared state. *Proceedings of the ACM SIGPLAN Commercial Users of Functional Programming, CUFP '10*. ACM, New York.

Knight, S., Nguyen, H., Falkner, N., Bowden, R., Roughan, M. (2011). The internet topology zoo. *IEEE Journal on Selected Areas in Communications*, 29(9), 1765–1775.

Kong, N. (2017). Design concept for a failover mechanism in distributed SDN controllers. Thesis, San José State University.

Koponen, T., Casado, M., Gude, N., Stribling, J., Poutievski, L., Google, M.Z., Ramanathan, R., NEC, Y.I., NEC, H.I., NEC, T.H. et al. (2010). Onix: A distributed control platform for large-scale production networks. *Proceedings of the 9th Conference on Operating Systems Design and Implementation*, California.

Kotronis, V., Gämperli, A., Dimitropoulos, X. (2015). Routing centralization across domains via SDN. *Computer Networks*, 92(P2), 227–239.

Kreutz, D., Ramos, F.M.V., Veríssimo, P., Rothenberg, C.E., Azodolmolky, S., Uhlig, S. (2015). Software-defined networking: A comprehensive survey. *Proceedings of the IEEE*, 103(1), 63.

Kulinski, D. and Burke, J. (2012). NDNVideo: Random-access live and pre-recorded streaming using NDN. Technical report [Online]. Available at: https://named-data.net/wp-content/uploads/TRstreaming.pdf.

Kulkarni, M., Goswami, B., Paulose, J. (2021). Experimenting with scalability of software defined networks using pyretic and frenetic. *Proceedings of the International Conference on Computing Science, Communication and Security*. Springer.

Kumar, S.P. (2016). Adaptive consistency protocols for replicated data in modern storage systems with a high degree of elasticity. Thesis, CNAM, Paris.

Kumar, A., Jain, S., Naik, U., Raghuraman, A., Kasinadhuni, N., Zermeno, E.C., Gunn, C.S., Ai, J., Carlin, B., Amarandei-Stavila, M. et al. (2015). BwE: Flexible, hierarchical bandwidth allocation for WAN distributed computing. *Proceedings of the 2015 ACM Conference on Special Interest Group on Data Communication, SIGCOMM '15*. London.

Lakshman, A. and Malik, P. (2010). Cassandra: A decentralized structured storage system. *SIGOPS Operating Systems Review*, 44(2), 35–40.

Lange, S., Gebert, S., Spoerhase, J., Rygielski, P., Zinner, T., Kounev, S., Tran-Gia, P. (2015a). Specialized heuristics for the controller placement problem in large scale SDN networks. *Proceedings of the 2015 27th International Teletraffic Congress*. doi: 10.1109/ITC.2015.32.

Lange, S., Gebert, S., Zinner, T., Tran-Gia, P., Hock, D., Jarschel, M., Hoffmann, M. (2015b). Heuristic approaches to the controller placement problem in large scale SDN networks. *IEEE Transactions on Network and Service Management*, 12(1), 4–17.

Lapeyrade, R., Bruyère, M., Owezarski, P. (2016). Openflow-based migration and management of the TouIX IXP. *Proceedings of the 2016 IEEE/IFIP Network Operations and Management Symposium, NOMS 2016*, Istanbul, 25–29 April.

Levin, D., Wundsam, A., Heller, B., Handigol, N., Feldmann, A. (2012). Logically centralized? State distribution trade-offs in software defined networks. *Proceedings of the First Workshop on Hot Topics in Software Defined Networks, HotSDN '12*.

Li, C., Porto, D., Clement, A., Gehrke, J., Preguiça, N., Rodrigues, R. (2012). Making geo-replicated systems fast as possible, consistent when necessary. *Proceedings of the Conference on Operating Systems Design and Implementation*.

Li, Y., Su, X., Riekki, J., Kanter, T., Rahmani, R. (2016). A SDN-based architecture for horizontal internet of things services. *Proceedings of the 2016 IEEE International Conference on Communications (ICC)*. doi: 10.1109/ICC.2016.7511053.

Liang, K., Zhao, L., Chu, X., Chen, H.H. (2017). An integrated architecture for software defined and virtualized radio access networks with fog computing. *IEEE Network*, 31(1), 80–87.

Lin, P., Bi, J., Wolff, S., Wang, Y., Xu, A., Chen, Z., Hu, H., Lin, Y. (2015). A west-east bridge based SDN inter-domain testbed. *IEEE Communications Magazine*, 53(2), 190–197.

Liu, Y., Hecker, A., Guerzoni, R., Despotovic, Z., Beker, S. (2015). On optimal hierarchical SDN. *Proceedings of the 2015 IEEE International Conference on Communications (ICC)*. doi: 10.1109/ICC.2015.7249178.

Mantas, A. and Ramos, F.M.V. (2016). Consistent and fault-tolerant SDN with unmodified switches. *CoRR*, abs/1602.04211.

Mantas, A. and Ramos, F.M.V. (2019). Rama: Controller fault tolerance in software-defined networking made practical [Online]. Available at: https://arxiv.org/ pdf/1902.01669.pdf.

McKeown, N., Anderson, T., Balakrishnan, H., Parulkar, G., Peterson, L., Rexford, J., Shenker, S., Turner, J. (2008). Openflow: Enabling innovation in campus networks. *SIGCOMM Computer Communication Review*, 38(2), 69–74.

Mellouk, A., Hoceini, S., Tran, H.A. (2013). *Quality of Experience for Multimedia: Application to Content Delivery Network Architecture*. ISTE Ltd, London and John Wiley & Sons, New York.

Michel, O., Bifulco, R., Rétvári, G., Schmid, S. (2021). The programmable data plane: Abstractions, architectures, algorithms, and applications. *ACM Computing Surveys (CSUR)*, 54(4), 1–36.

Michel, O. and Keller, E. (2017). SDN in wide-area networks: A survey. *Proceedings of the 2017 4th International Conference on Software Defined Systems (SDS)*. doi: 10.1109/SDS.2017.7939138.

Moazzeni, S., Khayyambashi, M.R., Movahhedinia, N., Callegati, F. (2018). On reliability improvement of software-defined networks. *Computer Networks*, 133, 195–211.

Monsanto, C., Reich, J., Foster, N., Rexford, J., Walker, D. (2013). Composing software-defined networks. *Proceedings of the 10th USENIX Conference on Networked Systems Design and Implementation, NSDI '13*. USENIX Association, Berkeley.

Montaño, M., Torres, R., Ludeña, P., Sandoval, F. (2021). IoT management analysis using SDN: Survey. In *Applied Technologies*, Botto-Tobar, M., Montes León, S., Camacho, O., Chávez, D., Torres-Carrión, P., Vizuete, M.Z. (eds). Springer International Publishing, Cham. doi: 10.1007/978-3-030-71503-845.

Morgan, H. (2015). AtlanticWave-SDX: A distributed intercontinental experimental software defined exchange for research and education networking. Press release, AtlanticWave-SDX, April.

Moshref, M., Bhargava, A., Gupta, A., Yu, M., Govindan, R. (2014). Flow-level state transition as a new switch primitive for SDN. *Proceedings of the 2014 ACM Conference on SIGCOMM, SIGCOMM '14*. Chicago. doi: 10.1145/2619239.2631439.

Mostafaei, H., Kumar, D., Lospoto, G., Chiesa, M., Di Battista, G. (2021). DeSI: A decentralized software-defined network architecture for internet exchange points. *IEEE Transactions on Network Science and Engineering*, 8(3), 2198–2212.

Müller, L.F., Oliveira, R.R., Luizelli, M.C., Gaspary, L.P., Barcellos, M.P. (2014). Survivor: An enhanced controller placement strategy for improving SDN survivability. *Proceedings of the 2014 IEEE Global Communications Conference*. doi: 10.1109/GLOCOM.2014.7037087.

Muqaddas, A.S., Bianco, A., Giaccone, P., Maier, G. (2016). Inter-controller traffic in ONOS clusters for SDN networks. *Proceedings of the 2016 IEEE International Conference on Communications (ICC)*. doi: 10.1109/ICC.2016.7511034.

MySQL (n.d.). Available at: http://www.mysql.fr/ [Accessed 2 March 2020].

Nguyen, T.D., Chiesa, M., Canini, M. (2017). Decentralized consistent updates in SDN. *Proceedings of the Symposium on SDN Research*, California.

Nguyen Ba, C.S. (2015). Adaptive control for availability and consistency in distributed key-values stores. PhD thesis, University of Illinois.

Obadia, M., Bouet, M., Leguay, J., Phemius, K., Iannone, L. (2014). Failover mechanisms for distributed SDN controllers. *Proceedings of the 2014 International Conference and Workshop on the Network of the Future (NOF)*, doi: 10.1109/ NOF.2014.7119795.

Ojo, M., Adami, D., Giordano, S. (2016). A SDN-IoT architecture with NFV implementation. *Proceedings of the 2016 IEEE Globecom Workshops (GC Workshops)*. doi: 10.1109/GLOCOMW.2016.7848825.

Oki, B. and Liskov, B. (1988). Viewstamped replication: A new primary copy method to support highly-available distributed systems. *Proceedings of the 7th Annual ACM Symposium on Principles of Distributed Computing (PODC)*. ACM.

Open Networking Foundation (2009). OpenFlow switch specification. Technical report, ONF. Available at: https://www. opennetworking.org/ [Accessed 27 January 2021].

Open Networking Foundation (2014). OF-CONFIG 1.2: OpenFlow Management and Configuration Protocol. Technical report, ONF [Accessed 5 January 2021].

Open Networking Foundation (n.d.). Available at: https://www. opennetworking.org/ [Accessed 19 January 2021].

Ongaro, D. and Ousterhout, J. (2014). In search of an understandable consensus algorithm. *Proceedings of the 2014 USENIX Annual Technical Conference, USENIX ATC '14* . USENIX Association, Berkeley.

ONOS (n.d.). Available at: https://onosproject.org/ [Accessed 2 January 2021].

OpenAI Gym Project (n.d.). Available at: https://gym.openai.com/ [Accessed 5 April 2019].

OpenConfig (n.d.). Available at: http://www.openconfig.net/ [Accessed 27 January 2021].

OpenDayLight Project (n.d.). Available at: http://www.opendaylight.org/ [Accessed 5 January 2021].

Open vSwitch (n.d.). Available at: http://www.openvswitch.org/ [Accessed 5 January 2020].

Oracle (n.d.). Available at: https://www.oracle.com [Accessed 24 October 2020].

Panda, A., Scott, C., Ghodsi, A., Koponen, T., Shenker, S. (2013). CAP for networks. *Proceedings of the 2nd ACM SIGCOMM Workshop on Hot Topics in Software Defined Networking, HotSDN '13*. ACM, New York.

Pashkov, V., Shalimov, A., Smeliansky, R. (2014). Controller failover for SDN enterprise networks. *Proceedings of the 2014 International Science and Technology Conference (Modern Networking Technologies) (MoNeTeC)*. doi: 10.1109/MoNeTeC.2014.6995594.

Perrot, N. and Reynaud, T. (2016). Optimal placement of controllers in a resilient SDN architecture. *Proceedings of the 12th International Conference on the Design of Reliable Communication Networks (DRCN)*. doi: 10.1109/DRCN.2016.7470849.

Phemius, K., Bouet, M., Leguay, J. (2013). DISCO: Distributed multi-domain SDN controllers. *CoRR*, abs/1308.6138.

Pontarelli, S., Bonola, M., Bianchi, G., Capone, A., Cascone, C. (2015). Stateful OpenFlow: Hardware proof of concept. *Proceedings of the 2015 IEEE 16th International Conference on High Performance Switching and Routing (HPSR)*, Budapest.

POX (n.d.). Available at: https://noxrepo.github.io/pox-doc/html/ [Accessed 22 April 2021].

Qiu, K., Huang, S., Xu, Q., Zhao, J., Wang, X., Secci, S. (2017). ParaCon: A parallel control plane for scaling up path computation in SDN. *IEEE Transactions on Network and Service Management*, PP(99), 1–1.

Rana, D.S., Dhondiyal, S.A., Chamoli, S.K. (2019). Software defined networking (SDN) challenges, issues and solution. *International Journal of Computer Sciences and Engineering*, 7(1), 884–889.

Reitblatt, M., Foster, N., Rexford, J., Walker, D. (2011). Consistent updates for software-defined networks: Change you can believe in! *Proceedings of the 10th ACM Workshop on Hot Topics in Networks*, New York.

Ren, W., Sun, Y., Luo, H., Guizani, M. (2019). A novel control plane optimization strategy for important nodes in SDN-IoT networks. *IEEE Internet of Things Journal*, 6(2), 3558–3571.

Ros, F.J. and Ruiz, P.M. (2016). On reliable controller placements in software-defined networks. *Computer Communications*, 77(C), 41–51.

Rothenberg, C.E., Nascimento, M.R., Salvador, M.R., Corrêa, C.N.A., Cunha deLucena, S., Raszuk, R. (2012). Revisiting routing control platforms with the eyes and muscles of software-defined networking. *Proceedings of the 1st Workshop on Hot Topics in Software Defined Networks, HotSDN '12*. ACM, New York.

Ruslan, R., Othman, N.B., Fuzi, M.F.M., Ghazali, N. (2020). Scalability analysis in mininet on software defined network using ONOS. *Proceedings of the 2020 Emerging Technology in Computing, Communication and Electronics (ETCCE) Conference*. IEEE.

Ryu SDN Framework (n.d.). Available at: https://ryu-sdn.org/ [Accessed 10 March 2021].

Sakic, E. and Kellerer, W. (2018). Impact of adaptive consistency on distributed SDN applications: An empirical study. *IEEE Journal on Selected Areas in Communications*, 13.

Sakic, E., Sardis, F., Guck, J.W., Kellerer, W. (2017). Towards adaptive state consistency in distributed SDN control plane. *Proceedings of the 2017 IEEE International Conference on Communications (ICC)*. doi: 10.1109/ICC.2017.7997164.

Sakic, E., Đerić, N., Kellerer, W. (2018). Morph: An adaptive framework for efficient and byzantine fault-tolerant SDN control plane. *IEEE Journal on Selected Areas in Communications*, 36, 2158–2174.

Samaan, N. and Karmouch, A. (2009). Towards autonomic network management: An analysis of current and future research directions. *IEEE Communications Surveys and Tutorials*, 11(3), 22–36.

Sandhya, Y.S. and Haribabu, K. (2017). A survey: Hybrid SDN. *Journal of Network and Computer Applications*, 100, 35–55.

Sanner, J.M., Hadjadj-Aoufi, Y., Ouzzif, M., Rubino, G. (2016). Hierarchical clustering for an efficient controllers' placement in software defined networks. *Proceedings of the 2016 Global Information Infrastructure and Networking Symposium (GIIS)*. doi: 10.1109/GIIS.2016.7814936.

Santos, M.A.S., Nunes, B.A.A., Obraczka, K., Turletti, T., de Oliveira, B.T., Margi, C.B. (2014). Decentralizing SDN's control plane. *Proceedings of the 39th IEEE Conference on Local Computer Networks, LCN 2014* , Edmonton, 8–11 September.

Sarmiento, D.E., Lebre, A., Nussbaum, L., Chari, A. (2021). Decentralized SDN control plane for a distributed Cloud-edge infrastructure: A survey. *IEEE Communications Surveys & Tutorials*, 23(1), 256–281.

Schiff, L., Schmid, S., Kuznetsov, P. (2016). In-band synchronization for distributed SDN control planes. *SIGCOMM Computer Communication Review*, 46(1), 37–43.

Schütz, G. and Martins, J.A. (2020). A comprehensive approach for optimizing controller placement in software-defined networks. *Computer Communications*, 159, 198–205.

Shalimov, A., Zuikov, D., Zimarina, D., Pashkov, V., Smeliansky, R. (2013). Advanced study of SDN/openflow controllers. *Proceedings of the 9th Central & Eastern European Software Engineering Conference in Russia, CEE-SECR '13*. ACM, New York.

Shamshirband, S., Fathi, M., Chronopoulos, A.T., Montieri, A., Palumbo, F., Pescapè, A. (2020). Computational intelligence intrusion detection techniques in mobile cloud computing environments: Review, taxonomy, and open research issues. *Journal of Information Security and Applications*, 55, 102582.

Sharkh, M.A., Jammal, M., Shami, A., Ouda, A. (2013). Resource allocation in a network-based cloud computing environment: Design challenges. *IEEE Communications Magazine*, 51(11), 46–52.

Shin, S., Song, Y., Lee, T., Lee, S., Chung, J., Porras, P., Yegneswaran, V., Noh, J., Kang, B.B. (2014). Rosemary: A robust, secure, and high-performance network operating system. *Proceedings of the 2014 ACM SIGSAC Conference on Computer and Communications Security, CCS '14*. Scottsdale.

Shirmarz, A. and Ghaffari, A. (2021). Taxonomy of controller placement problem (CPP) optimization in software defined network (SDN): A survey. *Journal of Ambient Intelligence and Humanized Computing*, 1–26.

Shraddha, K. and Emmanuel, M. (2014). Review and comparative study of clustering techniques. *International Journal of Computer Science and Information Technology (IJCSIT)*, 5, 805–812.

Sinalgo (n.d.). Sinaglo – Simulator for Network Algorithms. Available at: https://sinalgo.github.io/ [Accessed 1 February 2021].

Singh, A.K., Maurya, S., Kumar, N., Srivastava, S. (2020). Heuristic approaches for the reliable SDN controller placement problem. *Transactions on Emerging Telecommunications Technologies*, 31(2), e3761.

Sivasubramanian, S. (2012). Amazon dynamoDB: A seamlessly scalable non-relational database service. *Proceedings of the 2012 ACM SIGMOD International Conference on Management of Data, SIGMOD '12*. ACM, New York.

Sminesh, C.N., Kanaga, E.G.M., Roy, A. (2019). Optimal multi-controller placement strategy in SD-WAN using modified density peak clustering. *IET Communications*, 13(20), 3509–3518.

Spalla, E.S., Mafioletti, D.R., Liberato, A.B., Ewald, G., Rothenberg, C.E., Camargos, L., Villaca, R.S., Martinello, M. (2016). Ar2c2: Actively replicated controllers for SDN resilient control plane. *Proceedings of the 2016 IEEE/IFIP Network Operations and Management Symposium*. doi: 10.1109/NOMS.2016.7502812.

Stribling, J., Sovran, Y., Zhang, I., Pretzer, X., Li, J., Kaashoek, M.F., Morris, R. (2009). Flexible, wide-area storage for distributed systems with wheelfs. *Proceedings of the 6th USENIX Symposium on Networked Systems Design and Implementation, NSDI 2009*, Boston, 22–24 April.

Stringer, J.P., Pemberton, D., Fu, Q., Lorier, C., Nelson, R., Bailey, J., Corrêa, C.N.A., Rothenberg, C.E. (2014). Cardigan: SDN distributed routing fabric going live at an internet exchange. *Proceedings of the IEEE Symposium on Computers and Communications, ISCC 2014* , Funchal, 23–26 June.

Tavakoli, A., Casado, M., Koponen, T., Shenker, S. (2009). Applying NOX to the datacenter. *Proceedings of the Workshop on Hot Topics in Networks (HotNets-VIII)*, New York.

Tennenhouse, D.L. and Wetherall, D.J. (2007). Towards an active network architecture. *SIGCOMM Computer Communication Review*, 37(5), 81–94.

Terry, D.B., Prabhakaran, V., Kotla, R., Balakrishnan, M., Aguilera, M.K., Abu-Libdeh, H. (2013). Consistency-based service level agreements for cloud storage. *Proceedings of the 24th ACM Symposium on Operating Systems Principles, SOSP '13*. ACM, Farminton.

Tootoonchian, A. and Ganjali, Y. (2010). Hyperflow: A distributed control plane for openflow. *Proceedings of the 2010 Internet Network Management Conference on Research on Enterprise Networking, INM/WREN '10*. Berkeley.

Tootoonchian, A., Gorbunov, S., Ganjali, Y., Casado, M., Sherwood, R. (2012). On controller performance in software-defined networks. *Proceedings of the 2nd USENIX Conference on Hot Topics in Management of Internet, Cloud, and Enterprise Networks and Services, Hot-ICE '12*. Berkeley.

Tran, H.-A., Souihi, S., Tran, D.A., Mellouk, A. (2019). Mabrese: A new server selection method for smart SDN-based CDN architecture. *IEEE Communications Letters*, 23, 1012–1015.

Ul Huque, M.T.I., Si, W., Jourjon, G., Gramoli, V. (2017). Large-scale dynamic controller placement. *IEEE Transactions on Network and Service Management*, 14(1), 63–76.

Voellmy, A., Kim, H., Feamster, N. (2012). Procera: A language for high-level reactive network control. *Proceedings of the 1st Workshop on Hot Topics in Software Defined Networks, HotSDN '12*. ACM, New York.

Voldemort Project (n.d.). Design. Available at: http://www.project-voldemort.com/voldemort/design.html [Accessed 11 April 2016].

Wallin, S. and Wikström, C. (2011). Automating network and service configuration using NETCONF and YANG. *Proceedings of the 25th International Conference on Large Installation System Administration, LISA '11*. Berkeley.

Wang, G., Zhao, Y., Huang, J., Wu, Y. (2018). An effective approach to controller placement in software defined wide area networks. *IEEE Transactions on Network and Service Management*, 15(1), 344–355.

Wu, Y., Zhou, S., Wei, Y., Leng, S. (2020). Deep reinforcement learning for controller placement in software defined network. *Proceedings of the IEEE Conference on Computer Communications Workshops (INFOCOM WKSHPS)*. IEEE.

Yang, J., Yang, X., Zhou, Z., Wu, X., Benson, T., Hu, C. (2016). Focus: Function offloading from a controller to utilize switch power. *Proceedings of the 2016 IEEE Conference on Network Function Virtualization and Software Defined Networks (NFV-SDN)*, California.

Yap, K.-K., Motiwala, M., Rahe, J., Padgett, S., Holliman, M., Baldus, G., Hines, M., Kim, T., Narayanan, A., Jain, A. et al. (2017). Taking the edge off with Espresso: Scale, reliability and programmability for global internet peering. *Proceedings of the Conference of the ACM Special Interest Group on Data Communication, SIGCOMM '17*. ACM, New York.

Yazici, V., Sunay, M.O., Ercan, A.O. (2014). Controlling a software-defined network via distributed controllers. *CoRR*, abs/1401.7651.

Yeganeh, S.H. and Ganjali, Y. (2016). Beehive: Simple distributed programming in software-defined networks. *Proceedings of the Symposium on SDN Research, SOSR '16*. Santa Clara.

Yeganeh, S.H., Tootoonchian, A., Ganjali, Y. (2013). On scalability of software-defined networking. *IEEE Communications Magazine*, 51(2), 136–141.

Yu, H. and Vahdat, A. (2000). Design and evaluation of a continuous consistency model for replicated services. *Proceedings of the 4th Conference on Symposium on Operating System Design & Implementation – Volume 4, OSDI '00*. Berkeley.

Yu, M., Rexford, J., Freedman, M.J., Wang, J. (2010). Scalable flow-based networking with difane. *Proceedings of the ACM SIGCOMM 2010 Conference, SIGCOMM '10*. ACM, New York.

Yujie, R., Muqing, W., Yiming, C. (2020). An effective controller placement algorithm based on clustering in SDN. *Proceedings of the 2020 IEEE 6th International Conference on Computer and Communications (ICCC)*. IEEE.

Zhang, Q., Cheng, L., Boutaba, R. (2010). Cloud computing: State-of-the-art and research challenges. *Journal of Internet Services and Applications*, 1(1), 7–18.

Zhang, T., Bianco, A., Giaccone, P. (2016). The role of inter-controller traffic in SDN controllers placement. *Proceedings of the 2016 IEEE Conference on Network Function Virtualization and Software Defined Networks (IEEE NFV-SDN)*, California.

Zhang, B., Wang, X., Huang, M. (2018). Adaptive consistency strategy of multiple controllers in SDN. *IEEE Access*, 6, 78640–78649.

Zhang, X., Cui, L., Wei, K., Tso, F.P., Ji, Y., Jia, W. (2021). A survey on stateful data plane in software defined networks. *Computer Networks*, 184, 107597.

Zhong, J., Yates, R.D., Soljanin, E. (2018). Minimizing content staleness in dynamo-style replicated storage systems. *Proceedings of the IEEE Conference on Computer Communications Workshops (INFOCOM WKSHPS)*, New Jersey.

Index

Other titles from

in

Networks and Telecommunications

2022

BENAROUS Leila, BITAM Salim, MELLOUK Abdelhamid
Security in Vehicular Networks: Focus on Location and Identity Privacy
(New Generation Networks Set – Volume 1)

2021

LAUNAY Frédéric
NG-RAN and 5G-NR: 5G Radio Access Network and Radio Interference

2020

PUJOLLE Guy
Software Networks: Virtualization, SDN, 5G and Security (2nd edition
revised and updated)
(Advanced Network Set – Volume 1)

GONTRAND Christophe
Digital Communication Techniques

2019

LAUNEY Frédéric, PEREZ André
LTE Advanced Pro: Towards the 5G Mobile Network
Harmonic Concept and Applications

TOUNSI Wiem
Cyber-Vigilance and Digital Trust: Cyber Security in the Era of Cloud
Computing and IoT

2018

ANDIA Gianfranco, DURO Yvan, TEDJINI Smail
Non-linearities in Passive RFID Systems: Third Harmonic Concept and
Applications

BOUILLARD Anne, BOYER Marc, LE CORRONC Euriell
Deterministic Network Calculus: From Theory to Practical Implementation

LAUNAY Frédéric, PEREZ André
LTE Advanced Pro: Towards the 5G Mobile Network

PEREZ André
Wi-Fi Integration to the 4G Mobile Network

2017

BENSLAMA Malek, BENSLAMA Achour, ARIS Skander
Quantum Communications in New Telecommunications Systems

HILT Benoit, BERBINEAU Marion, VINEL Alexey, PIROVANO Alain
Networking Simulation for Intelligent Transportation Systems: High Mobile
Wireless Nodes

LESAS Anne-Marie, MIRANDA Serge
The Art and Science of NFC Programming
(Intellectual Technologies Set – Volume 3)

2016

AL AGHA Khaldoun, PUJOLLE Guy, ALI-YAHIYA Tara
Mobile and Wireless Networks
(Advanced Network Set – Volume 2)

BATTU Daniel
Communication Networks Economy

BENSLAMA Malek, BATATIA Hadj, MESSAI Abderraouf
Transitions from Digital Communications to Quantum Communications:
Concepts and Prospects

CHIASSERINI Carla Fabiana, GRIBAUDO Marco, MANINI Daniele
Analytical Modeling of Wireless Communication Systems
(Stochastic Models in Computer Science and Telecommunication Networks
Set – Volume 1)

EL FALLAH SEGHROUCHNI Amal, ISHIKAWA Fuyuki, HÉRAULT Laurent,
TOKUDA Hideyuki
Enablers for Smart Cities

PEREZ André
VoLTE and ViLTE

2015

BENSLAMA Malek, BATATIA Hadj, BOUCENNA Mohamed Lamine
Ad Hoc Networks Telecommunications and Game Theory

BENSLAMA Malek, KIAMOUCHE Wassila, BATATIA Hadj
Connections Management Strategies in Satellite Cellular Networks

BERTHOU Pascal, BAUDOIN Cédric, GAYRAUD Thierry, GINESTE Matthieu
Satellite and Terrestrial Hybrid Networks

CUADRA-SANCHEZ Antonio, ARACIL Javier
Traffic Anomaly Detection

LE RUYET Didier, PISCHELLA Mylène
Digital Communications 1: Source and Channel Coding

PEREZ André
LTE and LTE Advanced: 4G Network Radio Interface

PISCHELLA Mylène, LE RUYET Didier
Digital Communications 2: Digital Modulations

2014

ANJUM Bushra, PERROS Harry
Bandwidth Allocation for Video under Quality of Service Constraints

BATTU Daniel
New Telecom Networks: Enterprises and Security

BEN MAHMOUD Mohamed Slim, GUERBER Christophe, LARRIEU Nicolas,
PIROVANO Alain, RADZIK José
Aeronautical Air–Ground Data Link Communications

BITAM Salim, MELLOUK Abdelhamid
Bio-inspired Routing Protocols for Vehicular Ad-Hoc Networks

CAMPISTA Miguel Elias Mitre, RUBINSTEIN Marcelo Gonçalves
Advanced Routing Protocols for Wireless Networks

CHETTO Maryline
Real-time Systems Scheduling 1: Fundamentals
Real-time Systems Scheduling 2: Focuses

EXPOSITO Ernesto, DIOP Codé
Smart SOA Platforms in Cloud Computing Architectures

MELLOUK Abdelhamid, CUADRA-SANCHEZ Antonio
Quality of Experience Engineering for Customer Added Value Services

OTEAFY Sharief M.A., HASSANEIN Hossam S.
Dynamic Wireless Sensor Networks

PEREZ André
Network Security

PERRET Etienne
Radio Frequency Identification and Sensors: From RFID to Chipless RFID

REMY Jean-Gabriel, LETAMENDIA Charlotte
LTE Standards
LTE Services

TANWIR Savera, PERROS Harry
VBR Video Traffic Models

VAN METER Rodney
Quantum Networking

XIONG Kaiqi
Resource Optimization and Security for Cloud Services

2013

ASSING Dominique, CALÉ Stéphane
Mobile Access Safety: Beyond BYOD

BEN MAHMOUD Mohamed Slim, LARRIEU Nicolas, PIROVANO Alain
Risk Propagation Assessment for Network Security: Application to Airport Communication Network Design

BERTIN Emmanuel, CRESPI Noël
Architecture and Governance for Communication Services

BEYLOT André-Luc, LABIOD Houda
Vehicular Networks: Models and Algorithms

BRITO Gabriel M., VELLOSO Pedro Braconnot, MORAES Igor M.
Information-Centric Networks: A New Paradigm for the Internet

DEUFF Dominique, COSQUER Mathilde
User-Centered Agile Method

DUARTE Otto Carlos, PUJOLLE Guy
Virtual Networks: Pluralistic Approach for the Next Generation of Internet

FOWLER Scott A., MELLOUK Abdelhamid, YAMADA Naomi
LTE-Advanced DRX Mechanism for Power Saving

JOBERT Sébastien *et al.*
Synchronous Ethernet and IEEE 1588 in Telecoms: Next Generation Synchronization Networks

MELLOUK Abdelhamid, HOCEINI Said, TRAN Hai Anh
Quality-of-Experience for Multimedia: Application to Content Delivery Network Architecture

NAIT-SIDI-MOH Ahmed, BAKHOUYA Mohamed, GABER Jaafar, WACK Maxime
Geopositioning and Mobility

PEREZ André
Voice over LTE: EPS and IMS Networks

2012

AL AGHA Khaldoun
Network Coding

BOUCHET Olivier
Wireless Optical Communications

DECREUSEFOND Laurent, MOYAL Pascal
Stochastic Modeling and Analysis of Telecoms Networks

DUFOUR Jean-Yves
Intelligent Video Surveillance Systems

EXPOSITO Ernesto
Advanced Transport Protocols: Designing the Next Generation

JUMIRA Oswald, ZEADALLY Sherali
Energy Efficiency in Wireless Networks

KRIEF Francine
Green Networking

PEREZ André
Mobile Networks Architecture

2011

BONALD Thomas, FEUILLET Mathieu
Network Performance Analysis

CARBOU Romain, DIAZ Michel, EXPOSITO Ernesto, ROMAN Rodrigo
Digital Home Networking

CHABANNE Hervé, URIEN Pascal, SUSINI Jean-Ferdinand
RFID and the Internet of Things

GARDUNO David, DIAZ Michel
Communicating Systems with UML 2: Modeling and Analysis of Network Protocols

LAHEURTE Jean-Marc
Compact Antennas for Wireless Communications and Terminals: Theory and Design

PALICOT Jacques
Radio Engineering: From Software Radio to Cognitive Radio

PEREZ André
IP, Ethernet and MPLS Networks: Resource and Fault Management

RÉMY Jean-Gabriel, LETAMENDIA Charlotte
Home Area Networks and IPTV

TOUTAIN Laurent, MINABURO Ana
Local Networks and the Internet: From Protocols to Interconnection

2010

CHAOUCHI Hakima
The Internet of Things

FRIKHA Mounir
Ad Hoc Networks: Routing, QoS and Optimization

KRIEF Francine
Communicating Embedded Systems / Network Applications

2009

CHAOUCHI Hakima, MAKNAVICIUS Maryline
Wireless and Mobile Network Security

VIVIER Emmanuelle
Radio Resources Management in WiMAX

2008

CHADUC Jean-Marc, POGOREL Gérard
The Radio Spectrum

GAÏTI Dominique
Autonomic Networks

LABIOD Houda
Wireless Ad Hoc and Sensor Networks

LECOY Pierre
Fiber-optic Communications

MELLOUK Abdelhamid
*End-to-End Quality of Service Engineering in Next Generation
Heterogeneous Networks*

PAGANI Pascal *et al.*
Ultra-wideband Radio Propagation Channel

2007

BENSLIMANE Abderrahim
Multimedia Multicast on the Internet

PUJOLLE Guy
Management, Control and Evolution of IP Networks

SANCHEZ Javier, THIOUNE Mamadou
UMTS

VIVIER Guillaume
Reconfigurable Mobile Radio Systems

Printed and bound by CPI Group (UK) Ltd, Croydon, CR0 4YY

27/10/2024

14580248-0005